Understanding the Magic of the Bicycle

Basic scientific explanations to the two-wheeler's mysterious and fascinating behavior

Understanding the Magic of the Bicycle

Basic scientific explanations to the two-wheeler's mysterious and fascinating behavior

Joseph W Connolly
Physics/EE Department, University of Scranton

Morgan & Claypool Publishers

ISBN 978-1-6817-4441-4 (ebook)
ISBN 978-1-6817-4440-7 (print)
ISBN 978-1-6817-4443-8 (mobi)

DOI 10.1088/978-1-6817-4441-4

Version: 20161201

IOP Concise Physics
ISSN 2053-2571 (online)
ISSN 2054-7307 (print)

A Morgan & Claypool publication as part of IOP Concise Physics
Published by Morgan & Claypool Publishers, 40 Oak Drive, San Rafael, CA, 94903 USA

IOP Publishing, Temple Circus, Temple Way, Bristol BS1 6HG, UK

To my family of generations past and future—from grandparents who nudged and motivated a youngster with cheers of encouragement—to my parents, Mary and Joseph, for their guidance, hopes, and dreams that could not be ignored. To those of the present—children and grandchildren—for whom we offer our prayers and aspirations. Above all, to my wife Evelyn, who, a half century ago, began a journey down a path that only teenagers would not fear.

There is expectance here from both the sides.

Troilus and Cressida, iv, 5 William Shakespeare

Contents

Look back on time with kindly eyes,
He doubtless did his best;
How softly sinks his trembling sun
In human nature's west!

Emily Dickinson

Preface **xii**

Acknowledgements **xiv**

Author biography **xv**

1 Introduction–the magic of the wheel **1-1**

2 The evolution of the bicycle **2-1**

2.1 Beginnings 2-1
2.2 Baron Karl Von Drais's running machine 2-1
2.3 The boneshaker 2-4
2.4 Early refinements 2-6
2.4.1 Brakes 2-6
2.4.2 Angled fork 2-6
2.4.3 Freewheels 2-7
2.4.4 Tubular frame member 2-7
2.4.5 Wire spokes 2-8
2.4.6 Wheel sizes 2-8
2.5 High-wheelers 2-9
2.6 Further refinements 2-9
2.6.1 Tires 2-9
2.6.2 Geared propulsion 2-10
2.7 The safety bicycle 2-11
2.8 Pneumatic tires 2-12
2.9 Bearings 2-13
2.10 Rider position 2-13
2.11 Materials 2-14
References 2-15

3 A review of basic ideas 3-1

3.1 Algebra 3-2
3.2 Trigonometry 3-2
3.2.1 Slope of a hill 3-4
3.3 Vectors 3-5
3.3.1 Specifying direction 3-5
3.4 Head to tail method of vector arithmetic 3-6
3.5 Resolution into components 3-7
3.6 Units of measurement 3-8
3.7 Unit conversions 3-10
3.8 Density 3-11
3.9 Concepts of mass 3-12
3.10 Center of mass 3-13
3.11 Our standard rider 3-14

4 Linear motion 4-1

4.1 Kinematics—the study of motion 4-1
4.1.1 Instantaneous speed 4-2
4.1.2 Velocity 4-2
4.1.3 Acceleration 4-2
4.2 Headwinds and tailwinds 4-5
4.3 Riding uphill and downhill 4-6

5 Forces—Newton's laws of motion 5-1

5.1 Newton's First Law of Motion 5-3
5.2 Newton's Second Law of Motion 5-4
5.3 Units of force, motion, mass 5-5
5.4 Newton's Third Law of Motion 5-6
5.5 Role of arm muscles 5-8
5.6 Frictional forces—a simple model 5-9
5.7 Static and sliding friction 5-10
5.8 Friction as the propulsion force in walking 5-13
5.9 The acceleration and deceleration of the bicycle 5-14
5.9.1 Acceleration when starting from rest 5-15
5.9.2 Acceleration from an existing velocity 5-16
5.9.3 Deceleration from an existing velocity 5-16
5.10 Maximum acceleration of a bicycle 5-17

5.10.1 Maximum acceleration 5-17

5.10.2 Maximum deceleration 5-18

5.11 Velocity and acceleration of a bicycle 5-19

5.12 Resistive forces on a moving bicycle 5-20

5.13 Air resistance 5-21

5.14 Rolling resistance 5-25

5.15 Bearing resistance 5-27

5.16 Coasting—simplified analysis 5-29

5.17 Force analysis walking versus riding 5-30

5.18 Average versus instantaneous pedal force 5-32

References 5-33

6 Gravity 6-1

6.1 The basic physics of gravity 6-1

6.2 Weight of objects 6-2

6.3 'Weight' of object as measured by a scale 6-3

6.4 Force of gravity on a slope—the basic physics 6-4

6.5 Riding uphill at a constant speed 6-7

6.6 Terminal speed 6-8

6.7 Terminal speed coasting downhill on a bike 6-9

6.8 Personalized determination of resistive force parameters 6-11

7 Momentum–impulse 7-1

7.1 The basic physics of momentum 7-1

7.2 Momentum and Newton's Second Law 7-2

7.3 Impulse 7-2

7.4 Momentum and impulse aspects of bicycle accidents 7-3

8 Work–energy–power 8-1

8.1 Work 8-1

8.2 Kinetic energy 8-3

8.3 Frictional effects 8-5

8.4 Gravitational potential energy 8-6

8.5 Conservation of energy 8-7

8.6 Energy conversion between kinetic and potential on the bicycle 8-8

8.7 Power—the basic physics 8-9

8.8 Power and kinetic energy 8-12

8.9 Power output to overcome resistive forces on a bike 8-13
8.10 Efficiency considerations in muscular effort 8-15
8.11 Average speed versus average power 8-17
References 8-18

9 Temperature and heat 9-1

9.1 Temperature and its measurement 9-1
9.2 Heat 9-3
9.3 Units of heat 9-4
9.4 Heat generation on a bicycle 9-4
9.5 Mechanisms for heat transfer 9-5
9.6 Conduction 9-5
9.7 Convection 9-7
9.8 Radiation 9-9
9.9 Evaporation 9-11
9.10 Cooling effects of evaporation 9-14
9.11 Role of cycling clothing 9-15
9.12 Wind effects on cooling capacity of evaporation 9-15
9.13 Humidity and dew point 9-16
9.14 Specific heat 9-17
References 9-21

10 Rotational motion 10-1

10.1 Kinematics of circular motion 10-1
10.1.1 Angular velocity 10-4
10.1.2 Angular acceleration 10-4
10.2 Dynamics of circular motion 10-6
10.3 Rotational kinetic energy 10-8
10.4 Moment of inertia of non-point masses 10-9
10.5 Moment of inertia and rotational kinetic energy of bicycle wheel 10-10
10.6 Angular momentum 10-12
10.6.1 Direction of angular momentum 10-14
10.7 Role of angular momentum in a bicycle 10-14

11 Torque—applications to the bicycle 11-1

11.1 Basic physics of torque 11-1
11.2 Rotational equilibrium 11-2
11.3 Mechanical advantage 11-3

11.4	Energy aspects of a high mechanical advantage	11-5
11.5	Multiple lever system	11-5
11.6	Early direct drive bicycles	11-6
11.7	High-wheelers	11-7
11.8	The safety bicycle	11-8
11.9	Force transmission in a geared bicycle	11-9
11.10	Multispeed gearing—force analysis	11-11
11.11	Gearing and pedaling cadence	11-14
11.12	Gearing and pedaling force	11-15
11.13	Braking	11-19
11.14	Wheelies	11-23
11.15	Headers	11-25
	References	11-27

12	**Centripetal acceleration—turning and bicycle stability**	**12-1**
12.1	Review of Newton's laws—centripetal force and acceleration	12-1
12.2	Making a turn	12-4
12.3	Banked surface	12-6
12.4	Equilibrium and stability	12-6
12.5	Equilibrium and stability with multiple points of support	12-6
12.6	Stability of runners	12-8
12.7	Stability of sprinter	12-8
12.8	Equilibrium and stability with single point of support	12-9
12.9	Stability of broom when not in equilibrium	12-10
12.10	Stability of bicycle when not in equilibrium	12-12
12.11	Self stability of a bicycle	12-16
	12.11.1 Weight distribution of the handlebar–fork–front wheel	12-18
	12.11.2 Trail of the front wheel contact with the ground	12-19
12.12	Summation of bicycle stability	12-21
	Reference	12-22

Appendices

A	**Bibliography**	**A-1**
B	**Common unit conversions**	**B-1**
C	**Trigonometric values**	**C-1**

Preface

A short story about bikes and books

The kid was ten years old and craved a bike that, for reasons of family resources and parental worry, was unattainable. It would have been easier to get a Red Rider B-B gun. After winning at bingo four weeks in a row resulted in a pot of twenty dollars, Mom finally saw that a bicycle was in the stars; the worries became the matter of prayers. Even many decades ago, a nice new shiny Schwinn or three-speed English was beyond the twenty dollars. Pursuing newspaper want ads, the kid managed to find a two-wheeler for twelve dollars. It was an orange and white beauty—the rust gave it personality. After a few weeks of gleeful cruising areas of town never known to exist, it was time to spruce up the two-wheeler with a new coat of orange paint—two cans from the remaining eight dollars. Moderation is not a virtue to a ten-year-old kid with a can of spray paint—besides, the frame, wheels, pedals and spokes all looked beautiful in bright orange! The addition of a basket, horn, and light made the bike even more stunning. The other neighborhood kids agreed with envy.

With the outside dirt and rust gone, this bike must be just as dirty and rusty on the inside.

A visit to the gas station down the block and fifty cents yielded a couple of milk bottles filled with kerosene. With an old screwdriver and large pliers, it was not too hard to remove every moving part inside the bike. The kerosene overflowed as everything—various size bearings, cones, axels, washers—was simultaneously crammed into the milk bottles. It was amazing how many disks with tabs came out of that coaster brake. After a good soaking and wiping dry, it was time to spend another fifty cents for some grease and put a very large puzzle back together. While the brake had a lot of similar looking parts, most pieces found a home.

These leftover parts must be extras.

The best way to test everything was on a nice hill. The Orange Beauty climbed that hill with a quiet ease and grace; the ride down should have been even better. If the purpose of the brake was to stop the bike, it worked very, very well.

Talk about stopping on a dime!

The journey over the handlebars led to no serious bodily damage—kids grow up bouncing off the ground. The real damage was to the kid's heart—it was devastated. He now had an Orange Monster and years of bike thirst were crushed. The adults just shook their heads; no one had ever heard of a place that fixed bikes and besides, there was little left from those bingo winnings. Finally, Mom, as mothers always do, had a suggestion: 'Maybe there is a *book* in the little library, up the street, that will show you how to fix your bike'.

Sure, fat chance that a one-room library with 200 books would have one on how to fix bicycles!

Out of other options, what did the kid have to lose? As the saying goes 'it is better to be lucky than good'. Sure enough there it was—a book with clear instructions on how to put those 100 brake pieces back together. With help from the screwdriver and pliers, the Orange Beauty found its place once again in kid heaven.

And the kid learned something about books that made all the difference in his world.

After two more weeks of riding, the kid managed to crash the Beauty into a curb and crack off the top bar. It got fixed, but that is another story and lesson learned by the kid.

Man, that guy was crabby and I even paid him the 25 cents in cash! Next time, maybe I will look for a book on welding.

Acknowledgements

There were many, many who helped the author. Especially in the early days, Mr Jerry Flynn, who turned around the grades of a 'D' student with his explanations of the head to tail addition of vector components. Dr Eugene McGinnis, who shared both an unsurpassed physical insight and humanity. Mrs Nancy Laffey, who, in what must have been a fit of madness, offered to type the initial manuscript—200 pages of pencil scrawlings. My students, who challenged me and kept me young at heart. And, finally to Evelyn for her quiet encouragement and patient, sharp eyed proofing and editing.

Author biography

Joseph W Connolly

The author, Joseph W Connolly, is a Professor of Physics/EE at the University of Scranton. He has a BS degree from the University of Scranton, a MS degree from the University of Illinois and a PhD from the Pennsylvania State University. In a teaching career spanning five decades, he has taught close to four dozen different courses, many tailored for the non-science major. He served in the United States Army, Signal Corps, with an honorable discharge as a Captain. Other professional activities include several years in industry and two decades of industrial consulting in computer aided design and digital image processing.

At present, his spare time is occupied playing with grandchildren, woodworking and riding a tandem with his wife, Evelyn.

Chapter 1

Introduction–the magic of the wheel

Wherever the spirit wished to go, there the wheels went, and they were raised together with the living creatures; for the spirit of the living creatures was in the wheels.

New American Bible—Book of Ezekiel, chapter 1, verse 20

Where do we begin our search for an understanding of the unique and powerful attraction of the wheel? Why does this round form, a shape that can be rolled and moved with ease, appeal to an almost primordial instinct deep within the soul? Does the allure of a round figure originate with the Greek metaphysical concept of perfection of the circle? Does it go back hundreds of centuries earlier to the invention of the wheel? Early specimens of wheeled vehicles, dating back to the fourth millennium B.C., have been found in archeological digs in Mesopotamia and Europe

We offer for speculation that our innate human appreciation and fascination for round objects, capable of rolling with ease, may be found in human evolution of hundreds of thousands of years ago. Envision an early ancestor attempting to move rocks, building a shelter to protect against the wilds. Anyone who has tried to push a large rock across a yard quickly realizes that, rather than sliding the rock, it is easier to pry the boulder from one side and flip it over. The next realization is that a rock with a rounded shape will flip and roll with minimum effort. In a later chapter, we will see that the difference in effort is due to physical principles of sliding friction compared to rolling friction (resistance). Although our early hominid ancestors did not study such a chapter in their physics books, they surely would understand the ease with which they could move rounded objects.

doi:10.1088/978-1-6817-4441-4ch1

On their own axis as the planets run

Essay on man, iii, Alexander Pope

Yes—the origins of the bicycle may date back to our ancient ancestor's efforts in rolling stones!

A tantalizing clue to the human fascination with rolling objects is found in the etiology of the word *walk*. It derives from the Anglo-Saxon word 'wealcam', which means to move, revolve, or roll!

Another aspect of our primordial, instinctive appreciation of motion is that once we find an easier way to move objects—including our own body—we do not revert to the more difficult. The infant offers a fascinating example. As she struggles to move about, first by flipping, then by rolling, followed by crawling, she ultimately takes her first steps! There is no going back; at each 'step' there is little retrenchment. Once the baby takes her first few unsupported steps, crawling is no longer the way to go. The first steps bring boundless excitement to all—we can only wonder who has the greater joy—the little one or the parents (figure 1.1)!

We like to move, and we like to move as easily as possible. No doubt, much of the appeal of a bicycle is due to its ease of motion. It speaks to one of our deep, instinctive human desires.

The bicycle is a common, yet unique, mechanical contraption in our world. It is the most efficient mode of human powered transportation. Using the lower body muscles, a bicycle affords a several-fold gain in efficiency when compared to walking. Notwithstanding their utilitarian nature, bicycles are fun—they bring joy and exhilaration to the rider; there is a sense of amazement by observers.

In our world of electronic games and battery-powered gadgets, the bicycle is a classic toy still capable of delighting young children. Learning to balance and ride a two-wheeler is one of childhood's most satisfying memories. There is an experience of magic in the balancing.

Figure 1.1. First Steps, after Millet (1890), Vincent Van Gogh. Courtesy of Metropolitan Museum of Art, www.metmuseum.org.

Bicycles excite both youngsters and adults who are 'young at heart'. Once mastered as a child, the joy of riding a bicycle is never forgotten. The old adage 'once you learn to ride a bicycle, you never forget' is indeed true. Grownups are not immune to the exhilaration and sheer fun of a bicycle. Adults, who have not ventured onto a bike for decades, can leap upon a bike, pedal away, and feel ten years old again! Even the most unathletic and awkward feel a sense of grace and power on a moving two-wheeler.

A traveling bike is visual poetry; merely watching a passing cyclist evokes a sense of the transcendental.

In the late 1860s as bicycles appeared in major cities of the United States and Europe, the excitement and attraction of the self-propelled machines were captured in drawings and paintings by the most famous artists.

Early in his career, the American artist, Winslow Homer, worked as an illustrator for the legendary *Harper's Weekly* magazine. To Homer, the bicycle was a symbol of change. In his dramatization for the cover of the January 1, 1869 issue, it was 'out with the old in a wheelbarrow and in with the new on a bicycle' (figure 1.2).

In the latter half of the nineteenth century, bicycle crazes swept across Europe and the United States. These new-fangled contraptions, magically balanced on only two wheels, stunned observers. A cyclist, gliding by pedestrians, invoked responses ranging from awe to hostility. A century and a half later, some things have not changed. There is still awe in the magic of balancing. These machines continue to be a joy to ride but, unfortunately, continue to invoke a certain amount of hostility.

The self-propelled machines captivated Mark Twain, arguably the finest nineteenth century American writer. He wrote of his adventures learning to ride the bicycle in an essay *Taming the Bicycle*. Subsequently, Twain in his 1889 novel *A Connecticut Yankee in King Arthur's Court* saw the bicycle as a metaphor for the horse of the charging knights (figure 1.3).

Figure 1.2. *Harpers Weekly* Winslow Homer January 1, 1869.

KNIGHTS PRACTISING ON THE QUIET

Figure 1.3. A Connecticut Yankee in King Arthur's Court, Mark Twain [1].

It was on the 10th day of May 1884 that I confessed to age by mounting spectacles for the first time, and in the same hour I renewed my youth, to outward appearance, by mounting a bicycle for the first time … The spectacles stayed on.
speech by Mark Twain

The bicycle represents many things to many people. To Claude Monet the self-propelled wheeled vehicle was a toy for Jean, his young son (figure 1.4).

Figure 1.4. Jean Monet (1867–1913) on His Hobby Horse (1872), Claude Monet. Courtesy of Metropolitan Museum of Art, www.metmuseum.org.

Learning to balance and ride a bicycle is one of childhood's most lasting memories. The child experiences a sense of freedom and exhilaration that compares to the excitement when a baby takes his first steps. We are too young to recall the thrill of our first steps, whereas mastering the two-wheeler is a milestone never forgotten.

The Impressionist, Camille Pissarro saw the bicycle as an integral part of nineteenth century urban Paris (figure 1.5). The enlargement of the painting's lower region shows many enjoying a bicycle ride on this beautiful spring morning. Pissarro incorporated bicycle imagery in a number of his other paintings.

Another artist, Toulouse-Lautrec, captured the power and energy associated with the moving bicycle in a painting of a velodrome, an arena for fast bicycle racing (figure 1.6). The artist represents the racing cyclist as a blur in the background. There is no doubt that the ability to 'go fast' is one of the bicycle's attractions. For better or worse, it is very easy to attain high speed on a bicycle, unmatched in any other form of human powered locomotion. Even the most unathletic possesses a sense of power and grace while coasting effortlessly on a bicycle.

Figure 1.5. The Garden of the Tuileries on a Spring Morning (1899), Camille Pissarro. Courtesy of Metropolitan Museum of Art, www.metmuseum.org.

Figure 1.6. Tristan Bernard at the Velodrome Buffalo (1895), Toulouse-Lautrec. Private Collection.

For some, the bicycle is associated with images of beauty and social engagement by the upper class. The cycle was the heart of large social engagements. Well to do members of high society embraced the machines. Figure 1.7 captures the refined and genteel nature of the cycling meets.

Figure 1.7. Fashion of the Hour. Courtesy of Copake New York Antique Bicycle Auction.

Women, enjoying the freedom of independent travel, were especially drawn toward the bicycle. The appearance of bloomers in the female riders' dress was a matter of some outrage. Keen observers saw the bicycle playing a major role in the early stages of the suffrage movement. Stephan Crane, best known for his classic novel *The Red Badge of Courage*, perceived the bicycle foreshadowing profound changes in society.

> *Still, a second look at the Boulevard convinces one that the world is slowly, solemnly, inevitably coming to bloomers. We are about to enter an age of bloomers, and the bicycle, that machine which has gained economic position of the most tremendous importance, is going to be responsible for more than the bruises on the departed fat policeman of the Boulevard.*
>
> New York's Bicycle Speedway 1896 by Stephan Crane

Children quickly appreciate the freedom of independent transportation. At times, the wheeled toy encourages a child's spirit to escape the bounds of well-intended parental restrictions. The face of the young lady in Maurice Prendergast's watercolor shows a determination to use the adult distraction for a brief moment of freedom (figure 1.8).

Even the adults find the need to escape every now and then. There are days when a bicycle ride offers all a rare opportunity for solitude and peaceful reflection away from stress and pressure. Many of life's daily tribulations can, at least for a few hours, be left behind with a peaceful glide on two wheels. Some worries cannot keep pace at twenty miles per hour (figure 1.9).

Figure 1.8. Large Boston Public Garden Sketchbook: A Girl Riding a Tricycle in the Park (1895–97), Maurice Brazil Prendergast. Courtesy of Metropolitan Museum of Art, www.metmuseum.org.

Figure 1.9. Autumn Solitude. Courtesy of M. Mullen, The Times-Tribune.

Turn, turn my wheel! Turn round and round.
Without a pause, without a sound
So spins the flying world away.

Keramos by Henry Wadsworth Longfellow

The verse from Longfellow, composed in reference to the potter's wheel, conveys similar feelings to the cyclist enjoying an isolated ride.

Sadly, for many youngsters the bicycle was the unattainable – a first experience of intensely desiring something that, for one reason or another, could not be realized (figure 1.10).

This longing for a two-wheeler may have been the first time a child was motivated by the realization that a dream can be attained through persistence and hard work (figure 1.11).

Figure 1.10. Is your child left behind? David Robinson. Cycle Trades of America.

While the preceding images represent aspects of the bicycle that appeal to our human yearnings, no one has better captured the aspect of the bicycle that is the essence of this book than Norman Rockwell. To the perceptive illustrator of our life's most affecting moments, BICYCLES ARE FUN! (figure 1.12).

Yes, to Norman Rockwell, bicycles are REALLY FUN! (figure 1.13).

Sing, riding's a joy! For me I ride

Robert Browning

No matter what your age (figure 1.14). On a bicycle: you are 10 years old again! There is no doubt that a few turns of a pedal are the surest way to relive the joys of childhood. The bike is 'Rosebud' on two wheels!

Figure 1.11. New Departure Manufacturing Company advertisement.

Figure 1.12. School this year means more than ever before. Norman Rockwell.

Figure 1.13. Hey fellers! $100 in gold first prize. Norman Rockwell.

Figure 1.14. Hambidge Truth Magazine. Courtesy of Copake New York Antique Bicycle Auction.

The Child is the Father of the Man

The Rainbow by William Wordsworth

What makes this bicycle so much fun? Why do we get this childlike exhilaration while pedaling the two-wheeler? The bicycle, once mastered, bonds with our primordial desire to move, to move fast, and to move with ease. The bicycle extracts from our muscular efforts the maximum, most efficient form of human propulsion. Even when working hard, straining to push the machine against nature's forces of resistance, there is reward; the motion through the air affords a pleasant, cooling breeze. Another blessing is that, although we struggle mightily against the dominant resistive force of air resistance, the opponent is invisible, hidden from our senses. Somehow, this air resistance is a quiet adversary, almost imperceptible; it never seems to be oppressive.

Although cycling is viewed by most as a fun activity and almost everyone acquires the basic skills at a young age, few understand the laws of nature that give magic to the ride. Suppose we take a closer look at some of these fun, exhilarating, and magical aspects of cycling.

As the reader travels through this book, she will encounter fundamental explanations for a myriad of the bicycle's fascinating behavior. For example:

- A bicycle possesses an amazing ability to balance on two wheels. The balancing is seen as the result of a combination of motion toward the center of a turn and an inward lean (chapter 12).
- The self-stability of a bicycle, the way in which a falling bicycle picks itself back up, is a consequence of the tendency of objects to travel in a straight line (chapters 5 and 12).

- The need to lean during a turn, an instinctive action learned by the youngest of riders, is viewed as a necessary component of balancing and turning the two-wheeler (chapter 12).
- The cause of air resistance and the force's important significance in suppressing the forward motion of the bicycle is evaluated (chapter 5).
- The role of static friction, which normally opposes motion, is examined for its role in starting and stopping the cycle (chapter 5).
- The importance of gearing is considered for both the impact on the rider's effort and the machine's speed (chapter 11).
- The ultimate joy of freewheeling, being able to coast along with no effort, is examined as a consequence of the basic laws of motion (chapters 4 and 5).
- The geared transmission system is analyzed for its efficient conversion of the rotational motion of the pedals to the forward travel of the two-wheeler. We will learn how a magnification occurs—the wheels of the bike spinning as much as ten times faster than the feet (chapters 4, 10 and 11).
- The advantages of a modern bicycle's multigear system are explored. The rider has the benefit of selecting a high gear, suited for riding for speed, or a low gear that eases the challenge of climbing hills (chapters 8 and 11).
- Strenuous physical activity such as cycling requires large expenditures of energy at high power levels. We will examine the energetics of bicycling as a consequence of the need to overcome nature's resistive forces such as gravity, air resistance, rolling resistance, etc (chapters 5 and 8).
- A byproduct of the strenuous muscular efforts involved in the pedaling is the large amount of body heat generated. The various mechanisms of heat transfer are evaluated for their role in cooling the body (chapter 8 and 9).
- The cause of the wheel's rolling resistance and the mechanism by which a vertical deformation of the tire opposes the horizontal motion of the bicycle is considered (chapter 5).
- We examine the reasons for the bicycle offering the most efficient form of human powered locomotion—hint—it is not the gearing system (chapters 5 and 11).
- A two-wheeler steers with ease; the rider is able to control the machine's direction with small forces on the handlebars, sometimes just nudges or small leans (chapter 12).
- The moving cycle has a natural stability. It is astonishing how the bicycle, once in motion, is able to leave behind the intrinsic instability of a large mass perched high above two tiny points of support. Even without a rider, the moving bike wants to stay upright (chapter 12).
- The reader is asked to consider the ease with which a traveling bike can be stopped using the small appendages of our hands, especially when compared to the large lower body muscles needed to get the bicycle into motion (chapter 11).
- The delight experienced as the struggle of the climb is traded for the exhilaration of the downhill journey (chapter 6 and 8).
- The relationship between momentum and impulse is viewed as the mechanism through which helmets reduce collision forces on the head (chapter 7).

In spite of the bicycle's ubiquitous and fascinating presence in our everyday world, the bike's basic physical and mechanical principles are appreciated and understood by a select few. We invite the reader to join this privileged group. The mysterious behavior of these two-wheeled 'toys' can be understood as a consequence of simple physical principles.

A word of caution—it is common to find, in books and on the internet, explanations of the bicycle that are either erroneous or explained at very high mathematical levels. In this book, the reader will find simple, physically accurate explanations to a range of complex bicycle behavior. The approach is to start with fundamental principles, such as Newton's laws, and develop intuitive, basic explanations for the bicycle's behavior. Sprinkled throughout this book are a series of questions that inquire as to particular aspects of cycling. Each chapter will develop the answers to these questions.

But ... do you have to be a genius to understand both bicycles and physics? No, you do not have to be a genius to understand the physics of cycling. However, once you do understand the bike science—everyone will think you are a genius! (figure 1.15).

Figure 1.15. Dr Albert Einstein on a bicycle. Courtesy of the Archives, California Institute of Technology.

Thousands of geniuses live and die undiscovered—either by themselves or by others

Autobiography of Mark Twain

In the following pages, each concept is introduced and an elementary explanation offered based upon the simplest everyday example. Once the concept is established, we proceed to illustrate its application to the bicycle. For example, we first explore

Newton's laws as they apply to people pushing on a kitchen table; we then consider the application of Newton's laws to the bicycle. In a similar manner, the subtle ideas of static and sliding friction are examined as they pertain to a box on the floor. Literally, the next step is to consider how frictional forces enable walking. We subsequently detail the role of friction in the acceleration and deceleration of the bike. This approach is employed for many traditional concepts: motion, momentum, energy, power, rotation, temperature, heat, etc.

But isn't physics hard? Well, how about riding a bicycle? Don't we all remember what it was like when we first tried to balance the two-wheeler? At least we have never heard of anyone hitting a pothole or a tree and breaking a collarbone while thinking about the physical world!

We tried and tried; we did not give up. It took effort; it took practice but, eventually, with a little bit of help from a friend who knew: we did it; we got it right! It might not have been easy, but with practice, we got good at it. Once we learn the trick—riding a bike is easy and really, really fun!

When your own burden is heaviest, you can always lighten a little some other burden. At the times when you cannot see God, there is still open to you this sacred possibility, to show God; for it is the love and kindness of human hearts through which the divine reality comes home to men, whether they name it or not. Let this thought, then, stay with you: there may be times when you cannot find help, but there is no time when you cannot give help.

George S Merriam

That is how it is with science. It's like riding a bike! … and the best news is that we can use the fun of bicycling to learn the fun of the physical world.

The reader will find that many topics have application beyond the bicycle. For example:

- Any activity is subject to the principles of motion, force, energy, power, heat, and temperature. The fundamental laws of nature act on all sporting endeavors whether they are the quiet, competitive competitions such as curling or shuffleboard or high-energy engagements found in football or rugby.
- The mechanics of many moving machines—cars, motorcycles, skateboards etc, are similar to those of a bicycle. Even ice skating and sprinting share a common behavior with cycling.
- The concepts of air resistance, friction, and rolling resistance are very important to an automobile's behavior

Reference

[1] Twain M 1889 *A Connecticut Yankee in King Arthur's Court* (New York: Harper and Brothers) pp 364–5

Chapter 2

The evolution of the bicycle

How has the modern, 21st century bicycle evolved since it was first conceived in the early 19th century?

This chapter is a brief summary of the evolution of the two-wheeled human powered machine we call the bicycle. The basic principles and laws of physics are the foundation for the bicycle's successful development. For a comprehensive overview of early efforts to develop mechanical replacements for animals and the invention of the two-wheeler, the reader is invited to explore excellent monographs by Herlihy, Velox, JFB, and Ritchie (Appendix A Bibliography).

2.1 Beginnings

The use of animals for transportation originated in a prehistory of many millennia past. While horses allowed for travel over distances far greater and faster than could be achieved by walking and running, the animals were an expense and required substantial amounts of food and care. There was great interest for a contrivance that provided locomotion over long distances with less human effort and without the reliance on animals. As far back as the late seventeenth century, the French mathematician, Ozanam, offered a prize to the first person who devised a human-propelled machine to replace the horse. Early attempts were based on carriage designs equipped with pedals and levers driven by humans. These designs were overly complex and heavy to be propelled efficiently by humans. Trying to adapt a machine normally pulled by one or more horses, into some type of contraption suitable for a human to impel, is doomed to fail.

2.2 Baron Karl Von Drais's running machine

The machine considered as the true forerunner of the modern bicycle has an interesting and tragic origin. The triggering event leading to its development was the 1815 eruption of the volcano Tambora in Indonesia. This eruption was so powerful that its

doi:10.1088/978-1-6817-4441-4ch2

roar resounded hundreds of miles away. Its significance lay in the large quantities of dust pumped into the upper level of the atmosphere; the resulting red skies endured for years. Such large quantities of dust increase the reflection of sunlight, thereby, reducing the amount of solar energy that strikes the Earth's surface. Consequently, there was a reduction of global temperatures; the year after Tambora's eruption was called the 'year without a summer'. The year 1816 was noteworthy in history for its killing frosts and loss of food crops in the summer months in North America and Europe. The famine is blamed for tens of thousands of deaths. This massive food shortage led to a reduction in the number of horses. No doubt, the human population consumed the available grains and horses were viewed as a food source.

I had a dream, which was not all a dream.
The bright sun was extinguished, and the stars
Did wander darkling in the eternal space,
Rayless, and pathless, and the icy earth
Swung blind and blackening in the Moonless air;

Darkness written in 1816 by Lord George Byron

From this tragedy, we find in 1817, an individual who invented an alternative to equine powered transportation. He was the Baron Karl von Drais of Baden (now part of Germany) and his invention was called the Laufmaschine ('running machine'). As his machines were replicated in various countries, they were given different names: the 'Draisine' (English), the 'Draisienne' (French), the velocipede, and the dandy horse.

Von Drais's running machine was a two-wheeled vehicle on which the rider straddled a saddle. The device was propelled by skipping or skimming the feet along the ground. Turning handlebars steered the front wheel. The machine was reputed to have a top speed of 8–10 miles per hour. Examination of the front wheel and fork mechanism suggests that the device could be balanced with the feet off the ground, allowing for coasting on level and downhill surfaces. In chapter 12, we will see that a front wheel, free to turn, is essential to the two-wheeler's balancing mechanism (figure 2.1).

Figure 2.1. The Laufmaschine.

Turns by its handle

The Cyclops by Percy Shelley

Your fine elegant rascal, that can rise,
And stoop, almost together; like an arrow;
Shoot through the air as nimbly as a star;
Turn short as doth a swallow; and be here,
And there, and here, and yonder, all at once.

Volpone, iii Ben Jonson

The original Draisine design, while never immensely popular, persisted for about fifteen years. Oftentimes, the machines were greeted with derision, considered as toys for the privileged. Many were quick to point out the hazards to pedestrians.

Their earth-convulsing wheels affright the city.

Hellas by Percy Shelley

As late as 1830, French police and letter carriers made their rounds on these new-fangled contraptions. However, hills and rough roads limited the efficiency gain of the rolling machines. Harsh winter weather with snow and icy roads created another obvious hindrance. The cold steel and wooden machines offered no warmth and social comfort to an isolated, stranded traveler. No doubt, in these conditions, the four-legged beast was a superior mode of transportation.

Possibly, the concept of balancing on an inherently unstable contraption may have intimidated many. A primordial fear of falling likely inhibited the development of a machine that required keeping the feet off the ground for extended periods. Even on modern bicycles, many novice riders prefer to keep their saddles low so their feet can easily reach the ground. Pedals that clip onto the rider's shoes can invoke a feeling of helpless entrapment until the rider has mastered the unclipping motion.

Figure 2.2. Tricycle. Courtesy of Michelin North America, Inc. All rights reserved.

Nevertheless, the desire to develop a replacement for the horse and horse-driven carriage was not abandoned. For the next few decades, most development in human power transportation was devoted toward three- and four-wheeled vehicles. These tricycles and quadricycles imposed a substantial drawback of weight and mechanical complexity. The vintage image in figure 2.2 also clearly shows a fourth wheel, an outrigger, designed to prevent the contraption from toppling backward.

As I looked at the living creatures, I saw wheels on the ground, one beside each of the four living creatures.

New American Bible—Book of Ezekiel, chapter 1, verse15

2.3 The boneshaker

Nearly fifty years after Baron von Drais's original running machine, a genius came up with an invention that offered the possibility of effective human-powered transportation. The device employed minimal use of materials to enhance substantially the efficiency of the human muscles. Who then should get credit for the development of the first two-wheeled, foot-powered, rotary-cranked, steerable velocipede that evolved into the modern bicycle? There is much controversy in this regard. The author David Herlihy offered a detailed summary of the controversy in his book *Bicycle* [1]. He reports that the carriage mechanic, Pierre Lallement, testified that it was he who conceived the idea in 1862. Lallement claimed that his working model was completed in 1863 and demonstrated on the streets of Paris; unfortunately, Lallement secured no patent for his machine. In 1865, he moved to the United States with some of the velocipede's components.

The spring of 1866 is a significant time in the emergence of the bicycle. Now in the United States, Lallement reassembled his machine and rode it through the streets of New Haven, Connecticut. An investor, James Carroll, apparently appreciated the value of the two-wheeler and encouraged Lallement to patent his device. The application for patent protection was filed in April 1866. Although there was much later controversy as to who was the true inventor of the 'bicycle', the first 'bicycle' patent in the world was, indeed, issued to Lallement and Carroll in the United States on November 20, 1866. In the patent application, the new and useful claim was for the arrangement of two wheels with treadles and guiding handlebars. Lallement's machine could travel at speeds of eight miles per hour (figure 2.3).

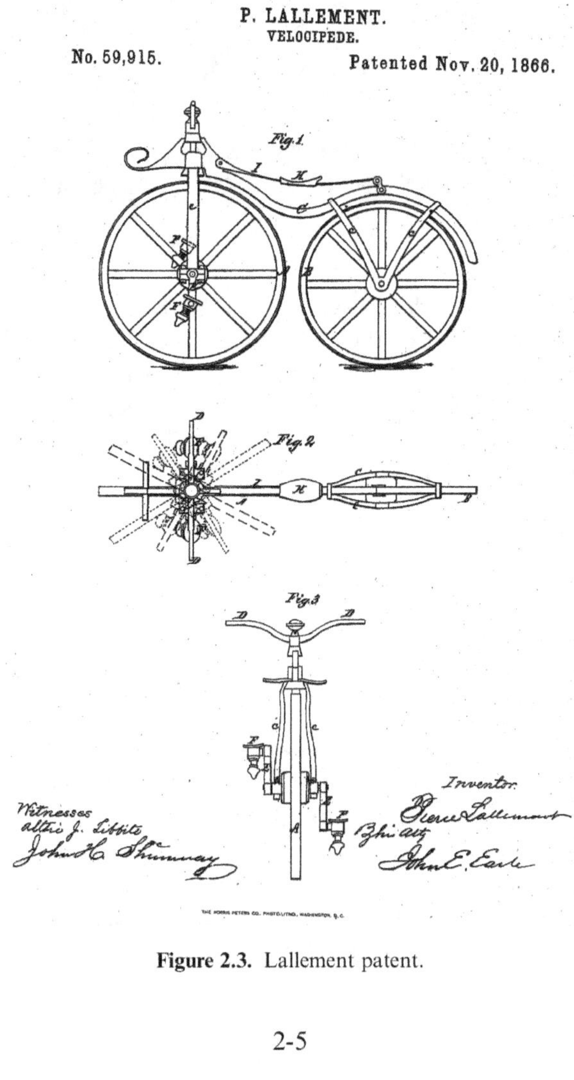

Figure 2.3. Lallement patent.

It can be argued that the original Draisine model and Lallement's improvements set a standard for efficient, minimal design that, in substance, has not been modified in the past one and a half centuries. The basic contrivance has only been improved in small, incremental steps. While each refinement has resulted in a better machine, the essence of the first velocipede persists.

A contending claim for the invention of the first bicycle was made later by the Parisian blacksmith, Pierre Michaux. After 1866, events developed rapidly on both sides of the Atlantic. Michaux began to manufacture velocipedes that bore a striking resemblance to the machine pictured in Lallement's patent application. His machines appeared in Paris during the spring and summer of 1867. The heavy wooden wheels were all too effective at transmitting rough road vibrations to the rider. The machines came to be known as the *boneshakers*.

2.4 Early refinements

Soon, after the first two-wheeler made its appearance, innovative thinkers began to change and improve the basic design.

2.4.1 Brakes

Examination of the machine's historical development shows that certain 'luxury' options such as brakes were found in the early Laufmaschines. Interestingly, children's bikes from the 1890s were produced without such an essential option. Even now, modern cycles such as track bikes and stunt bikes come without brakes.

The boneshaker brakes were usually applied to the rear wheels. An examination of the braking system on a variety of early bicycles reveals that designers sometimes employed front braking and sometimes rear braking. Complications in stopping the machine are explored in chapter 11. We will see that brake forces applied to the rear wheel are not as effective as when applied to the front wheel. Excessive front wheel braking creates the danger of heading over the handlebars.

2.4.2 Angled fork

A close examination of Lallement's patent reveals a machine with a near vertical front fork; however, by the late 1860s, velocipede designs began to show a headtube and fork that had a slight forward tilt. The angle is similar to a garden rake. Confusingly, this tilt is often called the fork rake or *rake angle*. In the boneshaker (figure 2.4), the angle is about 5° with the vertical. Most modern builders refer to this geometry as the *headtube angle*, measured from the horizontal; in figure 2.4, the angle from the horizontal is about 85°. It is suspected that the original reason for implementing the fork rake was to reduce the rider's reach to the handlebar.

Notwithstanding the requirement of a headtube construction robust enough to withstand the bending torque, there are several profound and compelling advantages to this tilted fork. Close examination of figure 2.4 shows that the angled fork causes the steering axis to strike the ground in front of the tire's point of ground contact. In this Michaux boneshaker, the steering axis strikes the ground approximately 1.5 inches in front of the tire's point of contact. The term *trail* is used to describe this

Figure 2.4. Michaux Boneshaker. Courtesy of Copake New York Antique Bicycle Auction.

parameter. In chapter 12, we examine how trail is an important parameter in determining the bicycle's steering characteristics.

The trail also plays a significant role in the *magical self-stability* of a bicycle. While Lallement's velocipede could be pedaled with the feet off the ground and balanced by steering the machine in the direction of the lean, its vertical fork provided no 'castering' effect to facilitate steering. Chapter 12 explores how an angled fork results in a force that encourages the bicycle to travel in a straight line.

2.4.3 Freewheels

Early bicycles had their pedal cranks directly attached to the front wheel. The pedals rotated as long as the bicycle was traveling. In order to rest the legs, it was necessary to remove the feet from the pedals and position them on stationary pegs attached to the fork. To resume pedaling, the rider had to 'catch' the revolving pedals. Innovators soon developed concepts for freewheels that allowed for 'coasting'—one of cycling's greatest pleasures. In his 1869 book, Velox offered a plan for a ratcheted axle enabling a free wheel design [2]. Velox's and other ratchet mechanisms did not catch on for many years.

2.4.4 Tubular frame member

Although, for the first few years, velocipede construction relied upon structural members fashioned from solid wood or iron, the original Michaux patent covered the use of hollow tubes [3]. By November 1868, racing bicycles from the New York firm of Pickering and Davis made use of tubular members in their refined boneshakers [4]. Tubular frames, because of their larger cross sectional diameter, offered the advantage of increased bending resistance when compared to solid bars of the same weight.

Figure 2.5. Bronze Chariot Inlaid with Ivory, Etruscan (6th Century B.C.). Courtesy of Metropolitan Museum of Art, www.metmuseum.org.

2.4.5 Wire spokes

By 1869, progress and improvements to the original boneshakers were rapid; a profound change occurred in spoke designs. The early velocipedes had wheels with heavy, thick spokes. Such designs may be traced to ancient applications of the wheel.

These clumsy, heavy spokes are even found in the wheels of elegant, bronze and ivory chariots from the first millennium B.C. Amazingly, these wheel designs persisted for thousands of years; it was the bicycle and the rider's desire to 'go fast' with minimal effort that brought about the radical change in the function of spokes (figure 2.5).

In the construction of traditional heavy wheels, large diameter spokes are compressed between the hub and the rim. An iron tire, heated and applied to the rim, held the entire structure in a state of compression. The compressive forces required thick spokes to avoid the problem of *buckling*. The result was a very heavy wheel. When horses supplied the power, they did not complain. However, with human power, the rider was keenly sensitive to the effort in spinning massive wheels. In chapter 10, we learn that the weight of the wheel is important due to the concept of rotational kinetic energy. Spokes, made of thin wires stretched between the hub and the rim, were a profound improvement. Tensioned wire spokes do not buckle; the new wheel designs were much lighter.

2.4.6 Wheel sizes

As with most technology, an improvement in one feature leads to changes and improvements in many other features. Such an evolution is spectacularly evident with the story of wire wheels. The velocipedes of the 1860s were propelled by pedals affixed directly to the front wheels; the distance traveled in one pedal rotation is the circumference of the wheel. The lightness of the wire wheel allowed designers to increase the diameter of the front wheel; thereby, increasing the distance traveled with each spin of the pedals. The improvement in the cycle's speed was immediate and dramatic.

Figure 2.6. Ordinary [5].

2.5 High-wheelers

In 1872, the 'speed geared ARIEL' made a brief appearance. It employed a small front wheel (34 inches) driven by a gear between the rotating crank and the wheel. However, the mechanical simplicity of merely increasing the diameter of the front wheel proved to be the most practical way to gain speed. The 1870s were the era of the *high-wheeler*. Many of our classic images of nineteenth century cycling are of these intriguing and eye catching machines. The mechanical marvels were sometimes disparagingly referred to as the 'penny farthing'. The disparity in the coins' size is a metaphor for the difference in the bicycle's front and rear wheels. The 'Ordinary' is a more favorable appellation for the high-wheeler.

The historical drawing in figure 2.6 clearly illustrates a number of significant details regarding the high-wheeler design. The rider had a lofty, forward position atop the front wheel. A careful inspection of figure 2.6 shows a front wheel brake located just forward of the headtube bottom; this brake is positioned to rub against the tire. Given the small load carried by the rear wheel, it was necessary to apply braking via the front wheel. The high-wheeler had a reputation for being extremely susceptible to 'headers' (chapter 11) over the handlebars from either hard braking or striking an obstruction.

In addition to speed increase afforded by the large wheels, the long wire spokes of the ordinaries offered improved isolation from the jolts of rough roads. Many high-wheeled bicycles also employed a suspended saddle held at the rear by a coiled spring.

2.6 Further refinements

2.6.1 Tires

And have the citizens gape at her and praise her tires.

The Alchemist, iv Ben Jonson

Since ancient times, when wheels made from wood were subject to rapid wear from rough roads, wheelsmiths covered the outside of the rim with a replaceable wear strip. Often, the wear surface was a leather band. Eventually, a more durable metal strip was utilized.

Wagon and carriage wheels of the nineteenth century still employed the old-fashioned technology. The common construction technique used a hub connected via heavy spokes to a wooden rim. An outside covering of a heated metal rim was placed around the outer perimeter. As this metal band cooled and contracted, it served a dual function holding the spokes in a state of compression and serving as a replaceable tire.

Such wagon wheel construction was adapted to velocipedes from the Draisine through the original boneshakers. To somewhat lessen the shaking of the bones, more refined wheels appeared in the late 1860s. A major improvement was the wheel from Price that held a solid rubber tire stretched and glued onto the rim. An illustration in the June 1869 issue of *Scientific American* shows wheels that were grooved to hold a rubber tire [6].

The rubber tires, in addition to softening the ride, afforded a better grip and stopping force when compared to their metal counterparts. The physics is that of the coefficients of friction (chapter 5). Subsequently, rubber tires were developed with a hollow core; a wire cable through the core secured the tire to the wheel. The next improvement in tire technology was the cushioned tire. Designed with a U shaped cross section, these tires further softened road vibrations.

2.6.2 Geared propulsion

In the high-wheeler, the inseam of the rider's leg limited the size of the front wheel. Thus, a turn of the crank would not carry shorter riders as far as their taller counterparts. The desire to achieve speeds that go beyond the high-wheel's size restriction and to avoid the danger of headers led to efforts to 'gear up' the bicycle. A significant machine was developed in 1878—*The Xtraordinary* by Messrs Singer and Company.

In figure 2.7, we find a photograph of a restored Xtraordinary offered at the Copake annual antique bicycle auction. A beautiful piece of machinery!

Figure 2.7. Singer Xtraordinary. Courtesy of Copake New York Antique Bicycle Auction.

This image shows the adoption of leveraged cranks to turn the front wheels. The leveraging of the drive mechanism effectively 'geared up' the front wheel; this gearing increases the drive wheel rotation per pedal cycle. By changing the attachment position of the levers, it was possible to vary the effective gear of the machine. Chapter 11 explores the advantages and disadvantages of leveraged mechanical systems. The pedal position also affords the rider a lower saddle position, thereby reducing the possibility of a header.

The lowered rider position has an additional advantage of a reduced profile through the air. In chapter 5, we consider the significance of air resistance on the moving rider/bicycle system; air resistance is the primary opposition force to the fast moving cycle.

While pedaling the Xtraordinary, the feet revolved in an oval motion.

2.7 The safety bicycle

Given the proclivity of a rider being pitched forward over the handlebars on bicycles with large front wheels, interest developed in returning to velocipedes with small wheels. These so-called 'safety' bicycles, with different size gears in the crank and rear wheel, could equal the speed of the high-wheelers. Chapter 11 investigates the principles of gearing up for speed and down for force.

Some early attempts at safety bicycles did not 'catch on'. However, in 1885, Messrs. Starley and Sutton introduced a bicycle that combined several concepts. This type of bicycle became the prototype for almost all future two-wheelers. The machine employed smaller front and rear wheels and was 'geared up' via a chain that connects rotary cranks and a driving rear wheel. The bicycle was named *The Rover Safety*. The 'Safety' appellation was intended to emphasize the desirability of riding closer to the ground and the cycle's relative immunity against headers. The saddle position was approximately midway between the front and rear wheels, resulting in better distribution of forces needed on the front wheel for steering and the rear wheel for propulsion. Chapters 11 and 12 discuss how to evaluate the ground forces acting at each wheel and the role played by the ground forces in accelerating, decelerating, and turning the bicycle.

I would give all of my fame for a pot of ale and safety.

Henry V, iii, 2 William Shakespeare

The initial implementation of the Rover Safety had a vertical steering fork. This required a system of coupling rods to connect the handlebars to the front wheel. The 'safety' aspect of using such a complex linkage to control the steering is debatable. Another complication for this design is that the saddle position and vertical fork caused the rider's feet to be very close to front wheel—creating interference during tight turns.

Within a few months, Starley and Sutton simplified the machine's design by substantially raking the front fork and allowing a seated rider to easily reach the handlebars directly joined to the fork (figure 2.8). The Starley machine received enthusiastic acceptance. The similarity to modern bicycle designs is a testament to

Figure 2.8. Rover Safety Sharp [7].

the soundness of the design. Many other prominent bicycle manufacturers produced similar bicycles. The days of the high-wheelers' popularity were ending.

The self-stability of a bicycle is one of its most magical aspects. A falling bicycle will tend to steer and right itself back up. This self-stability may be readily demonstrated even on a riderless bicycle. In chapter 12, we present a step-by-step, intuitive explanation for one of the bicycle's truly astonishing and mysterious behaviors.

2.8 Pneumatic tires

Invent some other tires!

Sun's Darling, iii, I, Dekker and Ford

While solid rubber tires and suspension mechanisms were somewhat effective in reducing road vibration, they carried a penalty of increased weight. The most dramatic improvement in the rideability of the bicycle is the development of air-filled tires.

The pneumatic tire was first invented in 1845 by the Scotsman, Robert W Thomson [8]. The date preceded the invention of the pedaled velocipede; Mr Thomson was too far ahead of his time. He envisioned his 'aerial' tires to be applied to carriage wheels and rocking chairs! We suspect the roads in the mid-1800s were a bit too rough for air-filled tires; although the rocking chair application still has a certain appeal. It was not until the last decade of the nineteenth century that James Dunlop of Dublin turned pneumatic tires into a commercial success. He was prodded into the development of the air-filled tire in response to his ten year old son's request to make his tricycle go faster!

Despite the inconvenience of pneumatic tire punctures, there are obvious benefits from their use:

- An improvement in comfort by affording a cushion against road bumps
- A reduction in energy loss from road bumps; the air tire acts like a spring absorbing impacts and then returning the energy. The rider's body is spared what would otherwise be an inelastic jolt.
- A low rolling resistance requiring less effort from the rider. The soft air tire is easy to compress compared to a solid rubber construction. In chapter 5, we

consider the phenomena and importance of rolling resistance. Professional racers were the first to embrace enthusiastically the new technology; they appreciated the wisdom of the ten-year-old kid. Within a few years, the pneumatic tire replaced the solid and cushioned rubber tires.

2.9 Bearings

Early bicycles used 'plain' bearings—merely an axle shaft constrained in a round bushing. As the shaft turned in its housing, the contact surfaces slid across one another. As moving surfaces rub, the phenomena of sliding friction causes major energy losses (chapter 5). Also, plain bearings required frequent lubrication and quickly wore. As the axle shaft erodes its cylindrical housing, the axle and, consequently, the wheel are subject to wobble.

A substantial reduction in frictional loss was attained by the invention of ball bearings by Jules-Pierre Suriray in 1869 [9]. How do these ball bearings work their magic? Why are they far superior to plain bearings? It is the *magic of the wheel*. As the shaft turns, the round bearings revolve within the housing; thus, we have a case of rolling resistance versus sliding friction. The physics of these resistive forces is explored in chapter 5.

2.10 Rider position

A variety of body positions may be assumed depending on the rider's preference and bicycle design. The rider on the left in figure 2.9 is in an upright position that presents a larger cross section to the air, whereas, on the right the rider is in a traditional bent over racing position. This racing position is more aerodynamic. In chapter 5, the physics of air resistance is explored. At typical riding speeds, air resistance is the primary force of opposition to the cyclist. Air resistance force is directly related to the rider's frontal cross section. In spite of his aerodynamic position, the rider in the racing stance might be well advised to consider less baggy clothes.

A second significant factor in riding stance is that the upright position results in a large fraction of the rider's weight being carried by the saddle. In the racing position, less weight is born by the saddle and; therefore, a larger portion of the rider's weight

Figure 2.9. Ride position [10].

is transferred onto the pedals. Thus, the downward pedal thrust is a combination of the rider's weight plus other muscular effort. The racing stance also enables the rider to gain additional thrust by pulling up on the handlebars. According to Newton's third law of motion (chapter 5), as the rider pulls up on the handlebars, the handlebars push down on him, thereby, adding to the total downward force on the pedals.

2.11 Materials

As patches set upon a little breach
Discredit more in hiding of the fault
Than did the fault before it was so patched.

King John, William Shakespeare

Von Drais's and Lallement's velocipedes used a top horizontal bar fashioned from wood into a serpentine shape. Michaux's early prototypes used a similar top bar. In his 1869 book, Velox recommended that the stout ash bar be formed by a wheelwright or coopersmith [11]. It would not be satisfactory to merely saw the shape from a large plank of wood; the wood grain would run out at the curves creating weak sections. A common technique to form a curved wooden bar is to steam bend in a form. The thick member would then require several days to dry while clamped in the form. Another method employs laminating wood strips that must be glued and clamped. Such fabrications result in a strong, relatively lightweight member. Properly fabricated wooden structural members have mechanical properties comparable to metal. The primary difficulty in wood construction is the labor and time intensive nature of the steam bending or lamination process. Wood species selection is also critical.

The wooden top bars were soon replaced with serpentine iron bars. The most expeditious technique would be to form the bars from cast iron; the geometry of the bar established by the mold. Cast iron is wholly unsuitable for structural components subject to loads. Velox described the use of cast iron as 'dangerous to the rider and pecuniary fatal to the manufacturer'. Cast iron has the propensity for unpredictable cracking failure. Michaux's early production of cast iron boneshakers quickly revealed the unsuitability of the brittle material [12]. The jolts and shocks created by riding the velocipedes on the rough roads of the era led to catastrophic failures. After only a year or so of using cast iron, the velocipede makers turned to a much more suitable material—wrought (worked) iron. Wrought iron is forged and then worked by a skilled blacksmith. The resulting material has a superior and more predictable response to stress.

The technology developments in bicycle materials have been at the crux of the two-wheeler's improvements for over a century. After using iron and steel, the manufacturers turned to aluminum, then titanium, and now carbon fiber—always with the goal of a lighter but strong machine (figure 2.10).

Figure 2.10. Aluminium Lightweight. Courtesy of Michelin North America, Inc.

References

[1] Herlihy D V 2004 *Bicycle* (New Haven, CT: Yale University Press) ch 4
[2] Velox 1869 *Velocipedes, Bicycles, and Tricycles* (New York: Routledge) p 95
[3] Roberts D 1991 *Cycling History Myths, and Queries—The Whole Blooming Truth* (Hadland Books) p 30
[4] Velox 1869 *Velocipedes, Bicycles, and Tricycles* (New York: Routledge) p 121
[5] Sharp A 1896 *Bicycles and Tricycles* (London: Longmans) (figure 122)
[6] Herlihy D V 2004 *Bicycle* (New Haven, CT: Yale University Press) p 125
[7] Sharp A 1896 *Bicycles and Tricycles* (London: Longmans) (figure 127)
[8] Thomson R W 1845 *UK patent* 10 990; 1847 *United States patent* 5104
[9] Herlihy D V 2004 *Bicycle* (New Haven, CT: Yale University Press) p 185
[10] Garratt H A 1899 *The Modern Safety Bicycle* (London: Whittaker and Co.) pp 180–1
[11] Velox 1869 *Velocipedes, Bicycles and Tricycles* (New York: Routledge) p 100
[12] Herlihy D V 2004 *Bicycle* (New Haven, CT: Yale University Press) p 93

Chapter 3

A review of basic ideas

You must crack the nuts before you can eat the kernel

Irish Proverb

This chapter is a review of several introductory, background topics. Most of the material was likely covered in a basic high school curriculum. Even if the reader does not recall everything from high school, this section will dust off a few cobwebs. If you really wish to understand the basic science of bicycles, you will need some familiarity with this chapter's content. It is important not to become discouraged and bogged down. You may wish to simply skim the chapter and come back to the details later. Topics, presented in this chapter, involve ideas and methods common throughout the book.

The math of physics involves basic concepts from algebra, trigonometry and vectors. Physics and math teachers are destined to go through life being told by almost everyone they meet how much that individual did not like math and science. The author remembers such a conversation as he sat for the first time in the chair of his new dentist. The thought occurred 'How dumb can this dentist be for telling a science professor how bad he was at math and science'. Then, in an instant, the next thought was which one of us was the dumber—the dentist or the patient who was about to allow the dentist to insert sharp, fast moving objects into his mouth. In fact, the dentist was very good at his profession and was quite talented with those sharp fast moving objects. He would have been better off keeping his inadequacies to himself.

It is better to remain silent and be thought a fool than to speak out and remove all doubt.

Attributed to President Abraham Lincoln, Mark Twain and others

The next few sections are a review and summary of basic math that is the language of physics.

doi:10.1088/978-1-6817-4441-4ch3

3.1 Algebra

Basic algebra offers the concise expression of fundamental physical principles. For instance, Newton's Second Law is commonly written as:

$$F = m\,a$$

where F is the force acting on the object, m is the object's mass, and a is the acceleration of the object. Do not worry about the specific meanings of these terms; they will be carefully defined and explored in chapter 5. For now, it is just a simple equation expressing a relationship between the terms.

If we know the force and the mass, we can algebraically rearrange the above equation to solve for the acceleration by dividing both sides of the above equation by m.

$$a = \frac{F}{m}$$

The above equations involve simple multiplication and division; on occasion, a bit more algebra is needed. For instance, the force of air resistance is very important to many cyclists. In chapter 5, we learn that the force of air resistance is proportional to the square of an object's speed v.

$$F \sim v\,v \sim v^2$$

The above is, strictly speaking, not an equation—it is a proportionality. The force does not equal the square of the speed; rather it is *proportional* to the square of the speed. If you double the speed, the force of air resistance increases by a factor of four ($2^2 = 4$). When your speed changes from 10 mph to 20 mph, the force of air resistance gets four times larger. It is very hard to pedal against the strong force of air!

Usually when we have proportionalities, it is useful to convert them into equations with proportionality factors. In chapter 5, we will see that the air resistance force also depends on parameters such as the frontal area of the bicycle and rider, the density of the air, etc. These parameters will be the proportionality factor.

3.2 Trigonometry

Now that we have brought back those algebra classes, is there other high school math to revisit? Yes—a little bit of trigonometry. Only a little bit but it is very, very important. Riding a bicycle in the real world involves forces and motion in three dimensions; we need some trig to understand the angles and relationships of the three dimensions. The basic ideas are described with a simple triangle. It is a special type of triangle—a *right triangle* (figure 3.1A). One of the three angles is 90°, a right angle; hence, the name *right triangle*.

Notice in this triangle that, in addition to the 90° angle, there are two other angles: θ *and* φ. Another little complication has cropped up; sometimes in science

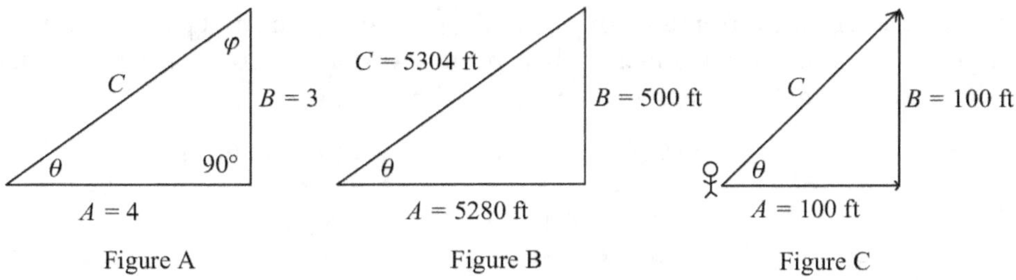

Figure 3.1. (A,B,C) Right triangle.

the Greek alphabet is used. Does that mean that the equations are going to get so complicated that the regular Latin/English alphabet of twenty-six lower case and twenty-six upper case characters will not be enough? No, not at all—the use of Greek letters is just another quirky thing that science does. Note, there are also three sides to the triangle—labeled as sides A, B, and C. As drawn, they look like (and often are) distances. The sides of the triangle can also represent forces, velocities, accelerations, etc.

This triangle with the three angles and three sides is labeled with a common nomenclature. The side opposite to the right angle is called the hypotenuse (side C). Looking at the angle θ, the side opposite to this angle is side B; the adjacent side to the angle θ is side A. There is a connection between A, B and C. It is called the *Pythagorean theorem.* The theorem, named for the Greek mathematician Pythagoras (570–495 B.C.), established a relation between the three sides of a right triangle: $C^2 = A^2 + B^2$

A triangle with $A = 4$, $B = 3$ has an hypotenuse:

$$C^2 = 4^2 + 3^2 = 16 + 9 = 25, \quad C = 5$$

Many craftsmen, such as carpenters, know the Pythagorean theorem. The author learned it when he was about eight years old. The teacher was the author's maternal grandfather who ended his formal education at the third grade in order to work in the coal mines of Northeastern Pennsylvania. If you are stuck with the Pythagorean theorem, seek out the nearest carpenter or mason.

Using the values for the sides of the triangle allows us to define the *trigonometric functions*; the three most common being the *sine*, the *cosine,* and the *tangent*. These functions are simply ratios of various sides of the right triangle. These are commonly written with the shorthand notation of sin, cos, and tan.

$$\tan\theta = \frac{\text{opposite}}{\text{adjacent}} = \frac{B}{A} = \frac{3}{4} = 0.75, \quad \sin\theta = \frac{\text{opposite}}{\text{hypothuse}} = \frac{B}{C} = \frac{3}{5} = 0.6, \quad \cos\theta = \frac{\text{adjacent}}{\text{hypothuse}} = \frac{B}{C} = \frac{4}{5} = 0.8$$

One other useful relationship is obtained if we write tangent as:

$$\tan\theta = \frac{\text{opposite/hypothuse}}{\text{adjacent/hypothuse}} = \frac{\sin\theta}{\cos\theta} \tag{3.1}$$

What is the use of these trig functions? Consider using a pencil and paper to draw the triangle of sides 3, 4, and 5 inches. If a protractor is used to measure the angles θ *and* φ, we will find the angles to be:

$$\theta = 36.9 \text{ degrees} \qquad \text{and} \qquad \varphi = 53.1 \text{ degrees}$$

We conclude that an angle of 36.9° has a sine of 0.600 and a cosine of 0.800. In a similar manner, we can determine that 53.1° has a sine of 0.800 and a cosine of 0.600. With additional effort, many more triangles could be drawn and a table constructed with all of the angles and associated sines, cosines, and tangents. Seems like a lot of work? Well, we do not have to do it—somebody already did. Appendix C is a tabulation of the common trig functions. Even better than appendix C is the use of a scientific calculator to obtain trig values. Trig functions are also commonly obtained online.

3.2.1 Slope of a hill

A fair question might be raised as to the purpose of this trig. Later chapters will have plenty of applications; hills are a significant aspect of the cycling experience. To cyclists the steepness of a hill is described by the term *slope*. The slope is expressed as a percentage of rise over run. Consider the hill shown in figure 3.1B; suppose you rise 500 ft and travel horizontally one mile long (5280 feet).

$$\text{slope} = \frac{\text{rise}}{\text{run}} \times 100\% = \frac{500 \text{ ft}}{5280 \text{ ft}} \times 100\% = 9.47\%$$

A tough hill to climb! In chapter 6, we learn that the angle or slope of the hill determines the fraction of the force of gravity that the rider must overcome. The angle of the hill is important; this angle is best obtained with the tangent function:

$$\tan\theta = \frac{500 \text{ ft}}{5280 \text{ ft}} = 0.095$$

From appendix C, the angle closest to this tangent value is between 5 and 6 degrees. If you have a 'scientific' calculator, you can skip using the table in appendix C. The calculator tells us the angle is about 5.4°. A useful relationship between the slope of a hill and its angle is:

$$\text{slope} = \tan\theta \times 100\% \tag{3.2}$$

For instance, if you use a carpenter's level and measure a hill to be 8°, the slope is:

$$\text{slope} = \tan 8 \times 100\% = 0.141 \times 100\% = 14.1\%$$

This hill offers another example of the Pythagorean theorem. Suppose we wish to compare to the reading of our bike's odometer, this is the distance C: $C = \sqrt{A^2 + B^2} = \sqrt{(5280 \text{ ft})^2 + (500 \text{ ft})^2} = 5304 \text{ ft}$.

Later, we examine the concept of *vector components*. It is proper to say that the hill, 5280 feet long, has a horizontal component of 5280 ft and a vertical component of 500 ft. This concept of components is extremely important in the world of physics.

Another example of the trigonometry of triangles is jaywalking. Everyone knows it is shorter to cut catty-corner across an intersection rather than cross one street, then the other. Suppose it is 100 ft East across the first street then another 100 ft North (figure 3.1C). What is the distance jaywalking across the intersection? From the Pythagorean theorem it will be: $C = \sqrt{(100\text{ ft})^2 + (100\text{ ft})^2} = 141$ ft. Therefore, by jaywalking, we travel 141 feet compared to 200 ft the proper way.

The above example is known to every kid as she cuts through yards on the way home from school. Amazingly, this concept has great implications! It is the basis of one of the most important mathematical concepts in the world of physics. We are referring to the concept of *vectors*! As the reader progresses through the various topics and chapters of the book, she might have to pause and scratch her head a bit. She might even encounter ideas that seem to violate common sense and intuition. Occasionally, it is a good idea to return to this section and think about cutting across people's yards on the way home from school.

3.3 Vectors

Back to our jaywalker (figure 3.1C)—her walk across the street can be called a *displacement*. A displacement is an example of a *vector*. A definition:

a vector is a physical quantity that has both magnitude and direction.

The direction of a vector is as equally important as the magnitude. As she crosses the street along A, her displacement vector is a magnitude of 100 ft and direction East. If she had gone 200 ft East, or 100 ft West, she would have ended up in a different spot. These would be different vectors.

To describe her journey across the streets using vector terminology: '100 ft, East' is her first displacement vector (vector **A**); '100 ft, North' is her second displacement vector (vector **B**). The girl's displacements are simple examples of vectors. Many other physical quantities are vectors, e.g. forces (10 lb, 'to the right'), velocity (25 mph, 'South'), acceleration, torques, moments (all defined in later chapters). Is everything in physics a vector? No, there are some physical quantities where there is no direction. We call these quantities *scalars*. An example of a scalar would be the mass of an object; temperature, energy, power, and density are other examples.

Another common concept is speed; speed is the magnitude of the velocity. If a bike is traveling 20 mph (no direction specified), we are using the scalar quantity of speed. If we say '20 mph to the right' (indicating a direction), we are referring to a vector quantity called velocity.

3.3.1 Specifying direction

Figure 3.1C, illustrates that cutting across the diagonal is a vector **C** equal to a distance of 141 ft in a Northeast direction. The use of compass points N, S, E, W or even the diagonals NE, SW etc. might often be adequate to describe a vector's direction. Eventually, the need for greater accuracy will entail the use of an *angular measure*, typically a 360° circle with the zero position at 3 o'clock. Counterclockwise is a positive sense; clockwise is a negative sense. Twelve o'clock is a +90° angle; the six o'clock position is 270° or a −90°. With this notation, the three vectors in figure 3.1C are described as:

A is 100 ft at an angle of 0° **B** is 100 ft at an angle of 90° **C** is 141 ft at an angle of 45°

What makes these vectors so special and important? There are several benefits—the first occurs when vectors are represented as arrows (as shown in figure 3.1C). Using an arrow to represent the vector, the length of the arrow is proportional to the magnitude of the vector, and the angle of the arrow is indicative of the direction of the vector. This graphical representation helps to visualize the physical quantity.

To construct the arrow, first establish a legend—a typical legend might be '1 inch of a drawing is 50 feet of displacement'. Thus, vector **A** would be 2 inches long with the arrow pointing to the right (figure 3.1C). A distance of 500 feet is an arrow 10 in long.

The next benefit of the arrow representation is that it provides a simple method to perform vector arithmetic such as addition and subtraction: *vector addition* employs a *head to tail method*. The description of the head to tail method sounds a bit wordy and complicated. The head of a vector is the pointy end of the arrow; the tail of the vector is the feathery end. Also, when the arrow is drawn on the paper, it is usually ok to slide its position around—as long as we do not change the length of the arrow or the direction of the arrow.

3.4 Head to tail method of vector arithmetic

In order to add two vectors (e.g. **A** + **B** = **C**), slide the second vector **B** and position its tail on the head of the first vector **A**. The sum of the two vectors is found by drawing a new vector **C** from the tail of the first to the head of the second. Although it is proper to call vector **C** the sum, we normally call it the *resultant vector*—more often called the *resultant*—(figure 3.1C).

The head to tail method gives a powerful, but very simple, technique for understanding how many types of vectors interact and combine with one another. In the above example, we illustrated the vector addition of displacement vectors. Many other physical quantities such as forces, velocities, accelerations, etc are also vectors. The head to tail method works for all types of vectors.

As another example: consider a situation in which two or more people exert a force on a table. A *force* is defined as a *push* or a *pull*. Suppose we have two people pushing a heavy table across the floor. Figure 3.2 is the view to a fly on the ceiling.

In figure 3.2A, if the first person pushes with a force of 120 lb (F_1) and the second person pushes in the same direction with a force of 40 lb (F_2)—what is the resultant force on the table? How do we add these forces? The two force vectors are drawn in

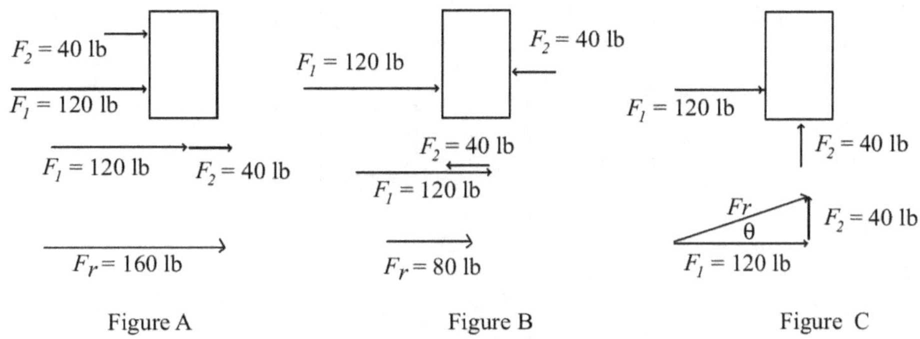

Figure 3.2. (A,B,C) Forces on table.

the head to tail manner; the resultant is seen as 160 lb. For clarity, the resultant has been displaced slightly downward.

What if the pushes were on opposite sides of the table? Figure 3.2B illustrates the force vectors and the head to tail addition. When we go from the tail of first vector to head of second vector, we see a resultant force of 80 lb to the right.

In these first two cases, we might have guessed the answer without using arrows and head to tail method. However, what if the first person pushes to the right (East) and the second person pushes North? The vector arrangement is shown in figure 3.2C. How do we get the value of the resultant F_r? One technique is to carefully draw the arrows according to some legend and then measure the length of the F_r arrow. A protractor gives the angle. This is called the graphical method. While it works, we can get a more accurate answer if we make use of our trigonometry skills.

The resultant magnitude is calculated from the Pythagorean theorem:

$$F_r^2 = (120 \text{ lb})^2 + (40 \text{ lb})^2, \quad F_r = 126 \text{ lb}$$

the angle is found by using the tangent and the functions in appendix C:

$$\tan\theta = \frac{40 \text{ ft}}{120 \text{ ft}} = 0.333; \quad \theta = 18.4°$$

Notice that the magnitude of the resultant force is about 126 lb, not much greater than the magnitude of F_1 alone. This is the nature of the geometric triangle; the person pushing North with 40 lb is adding very little to the overall effort.

What other kinds of vectors will we see in physics? As might be expected in a book on bicycles—we will encounter *velocity vectors.* A statement '20 mph at an angle of 25°' specifies a velocity vector. Other vectors are quantities such as acceleration, momentum, and torque.

3.5 Resolution into components

The above situation, adding a horizontal vector and a vertical vector into a resultant diagonal vector, can be thought of in a reversed way: any diagonal vector may be viewed as the sum of a horizontal vector and a vertical vector. The diagonal vector is made up of *horizontal and vertical components.* If someone exerts a single force on

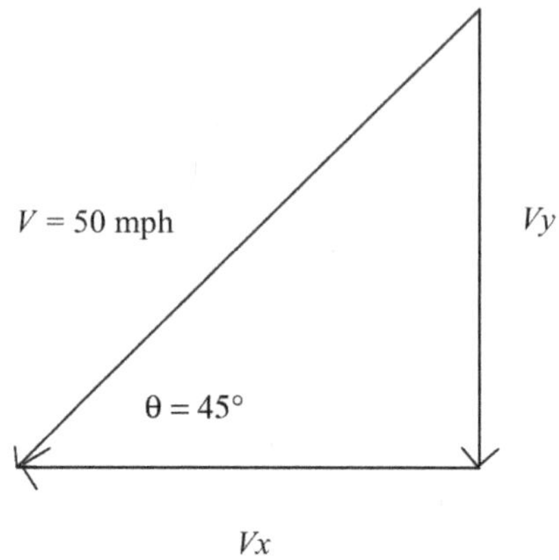

Figure 3.3. Velocity of car.

the table of 126 lb at an angle of 18.4°, it is equivalent to 120 lb to the right and 40 lb up. In other words, the 126 lb force has a horizontal component of 120 lb in the x-direction and a vertical component of 40 lb in the y-direction. The proof of this statement is seen in the head to tail drawing. When a vector is broken into its horizontal and vertical pieces, it is said to be 'resolved into components'.

Any diagonal vector may be resolved into horizontal and vertical components. This powerful concept will be explored in later chapters to comprehend fully its implications. For now, consider a few basic examples.

Think of a car traveling Southwest with a magnitude of 50 mph. What are the horizontal ($\mathbf{V}_x$ or West) and vertical ($\mathbf{V}_y$ or South) components of the car's velocity? The vector drawing is given in figure 3.3.

The nice thing about resolving into horizontal and vertical components is that you are guaranteed a right triangle; it is then a matter of using sines, cosines, and tangents. The car's velocity vector is the hypotenuse of the triangle.

The horizontal (x-component) is the side of the triangle opposite to the angle θ; hence, we use the sine of the angle θ: $\sin\ 45° = \dfrac{V_x}{50}$

$$V_x = 50 \text{ mph } \times \ \sin 45° = 50 \text{ mph } \times \ 0.707 = 35.4 \text{ mph}$$

The vertical component involves the cosine of the angle θ: $\cos 45° = \dfrac{V_y}{50}$

$$V_y = 50 \text{ mph } \times \ \cos 45° = 50 \text{ mph } \times 0.707 \ = 35.4 \text{ mph}$$

The original velocity of the car of 50 mph at an angle of 45 degrees has been resolved into an x-component of −35.4 mph and a y-component of −35.4 mph.

A vector's horizontal and vertical components are independent of one another.

3.6 Units of measurement

In this section we confront a frustrating and, frankly, unnecessary complication faced by many readers studying the physical world. The issue is *units of measurement*. Units indicate specific quantities—how tall you are, how much you weigh, how fast you are traveling, etc. The problem arises from a lack of logical consistency in the application of units. For example, forces are measured in pounds in the United States. The common units in the United States are called United States Customary.

If your weight is 120 lb, this means that the earth's gravity exerts a downward force on you of 120 lb. In the metric system, forces are measured in newtons. So what is your weight in newtons? You may be certain not too many people will know without taking a few seconds to convert. People from countries that use the metric system will state their bodyweight in kilograms—this is actually a body mass not a body weight. Another small complication—there are actually two flavors of the metric system. There was a time when they were referred to as the MKS (meter–kilogram–second) and the CGS (centimeter–gram–second) units. The MKS system is currently preferred and it is called the System International 'SI' system.

The world of units (or the units of the world) is a mess. Sadly, they complicate the life of a student trying to understand the physical world. When the author raises this issue with his fellow physics teachers, he often gets the response 'but the students will be better off knowing the metric system'. While it is true that students are better off knowing the metric system—we are all better off knowing a lot of things! However, we wonder how much the business of units interferes with the actual learning process in the study of physics. Maybe it is time to ask the art teachers to teach the metric system —no doubt, Renoir thought of his canvas sizes in centimeters. Alternatively, the track coach can do it—the future Olympians might have to compete in a 100-meter dash! At the very least, we should be consistent and use force units to specify a force such as weight, and mass units to specify masses. Perhaps, we might also eliminate terms like kilograms of force or kilometers per hour (the basic unit of time is the second).

Much of physics can be grasped at a fundamental level grounded on experience and intuition. When we talk to a student in the United States about a rock weighing 5 pounds or a car traveling at 25 miles per hour, she has a feeling for the size of the rock and the speed of the car. What instinctive feeling does she get for a rock weighing 107.8 newtons or a car traveling at 8.94 meters per second? Yet, physics is usually taught in the United States in these unfamiliar units!

To those who might think we are unfairly picking on the metric system, let us acknowledge the U.S. Customary units have their own set of quirks. *Mass* is measured in *slugs*; that alone sounds strange. We often mix in metric units—for example, calories for energy, watts to indicate power. We use inches, feet, yards, and miles for distance. We use ounces, pounds and tons for weight. We have two sizes of tons—short tons and long tons. Ounces are also units of volume along with pints, quarts, and gallons. Pressure is stated in pounds per square inch, sometimes—just pounds. Yes, they are a hodgepodge; however, these units are more familiar to the student than dynes, kilograms, meters, metric tons, bars, and kilopounds (in the world of units, this is the horse invented by the committee).

In this book, we will try our best to deal with concepts in familiar units. For the most part, the units will be U.S. Customary. In spite of the above rant, we acknowledge that there are times and situations when metric units are preferred. For some physical concepts, we will consider the quantity in several sets of units. For example, power is represented in foot–pounds per second, watts, horsepower, or

Table 3.1. Common units of measurement.

Quantity	U.S. Customary	Metric—SI, MKS	Metric, CGS
distance	feet (ft)	meter (m)	centimeter (cm)
	mile (mi)	kilometer (km)	
time	second (s)	second (s)	second (s)
	hour (h)	hour (h)	
force	pounds (lb)	newton (N)	dyne
mass	slugs	kilogram (kg)	gram(g)
energy	foot-pound (ft-lb/s)	joule (J)	erg
	calorie		
power	foot-pound/second (ft-lb/s)	watt	erg/second
	horsepower (hp)		

even Calories per hour. The calorie is a well-known unit for energy. Sadly, this world of units is a burden and an unfortunate complication to the joy of physics. It also causes physics teachers to go out on a bicycle ride to relieve their frustration—all are welcome to join in the ride.

For future reference, listed in table 3.1 are common units.

3.7 Unit conversions

The complexity caused by the various systems of units will be with us until the sun burns out. Even if the world suddenly adopted a uniform, consistent set of units, there would still be an extensive historical record. We best face the issue and learn how to cope with conversions between units. From time to time, we will have to convert a physical quantity from one set of units to another—for example: a speed in 'miles per hour' to a speed in 'feet per second'. We might also have to convert from one system to another—an air pressure in pounds per square inch (U.S. Customary) to an air pressure in bars (SI units).

Suppose you wish to convert a car's speed of 60 miles per hour into feet per second. The basic equivalencies are found in appendix B: 1 mile = 5280 feet, 1 hour = 60 minutes and 1 minute = 60 seconds. The confusion that develops is whether to multiply or divide by the equivalency. The simplest procedure is to set up a calculation in which the unit's *words* cancel according to the rules of basic algebra:

$$60\frac{\cancel{\text{mile}}}{\cancel{\text{hour}}} \times \frac{5280\ \text{feet}}{1\ \cancel{\text{mile}}} \times \frac{1\ \cancel{\text{hour}}}{60\ \cancel{\text{minute}}} \times \frac{1\ \cancel{\text{minute}}}{60\ \text{second}} = \frac{60 \times 5280}{60 \times 60}\frac{\text{feet}}{\text{second}} = 88\frac{\text{feet}}{\text{second}}$$

This particular conversion is especially helpful to remember: 60 mph = 88 ft/s; thus 30 mph = 44 ft/s, 120 mph = 176 ft/s, etc.

To convert the speed into SI metric of km/h, we use the equivalency: 1 mile = 1.61 km. The calculation is set up so that the unit words cancel:

$$\frac{60\ \cancel{\text{mile}}}{\text{hour}} \times \frac{1.61\ \text{km}}{\cancel{\text{mile}}} = \frac{60 \times 1.61}{1}\frac{\text{km}}{\text{hour}} = 96.6\frac{\text{km}}{\text{hour}}$$

3.8 Density

The notion of density arises in many physical instances. Density is intuitively easy to grasp; it is common in our everyday vocabulary. The weight density of a substance is defined by a basic expression in which w is the weight of the object and V is the volume; the Greek letter ρ (pronounced rho) is normally used to represent density:

$$\text{density} = \frac{\text{weight}}{\text{volume}} \text{ or as an equation } \rho = \frac{w}{V} \tag{3.3}$$

For instance, a cube of water one foot on a side will have a volume of one ft^3 and weighs 62.4 pounds. Thus, it has a density:

$$\rho = \frac{62.4 \text{ lb}}{1 \text{ ft}^3} = 62.4 \text{ lb/ft}^3$$

The above density is properly called the weight density. A mass density is written:

$$\text{density} = \frac{\text{mass}}{\text{volume}}$$

In addition to water, another substance that is extremely important to the cyclist is air—both on and off the bike. Not only do we breathe air, but also air is quite 'heavy' and must be pushed aside by the fast moving rider. Most of the exertion in riding a bike is caused by the resistance of the 'heavy' air.

Sometimes, we hear the expression that something is 'light as air'; suppose we consider how 'light' this really is. The density of air varies with temperature and atmospheric pressure. At a typical riding temperature of 60 °F and sea level pressure, air has a density of 0.0764 lb/ft^3. At less than one tenth of a pound per cubic foot, this does not seem very heavy! This issue requires further consideration. Imagine the biggest beach ball you could hold with your arms stretched wide—a nice red one, maybe four foot in diameter. Blow up the beach ball with just enough air to make it full but still soft. We do not want it to be under high pressure; high pressure represents extra air that has been crammed inside. How heavy is the air inside the beach ball?

An expression from fifth grade gives the volume of the sphere:

$$V = \frac{4}{3}\pi r^3 = \frac{4}{3}\pi(2 \text{ ft})^3 = 33.5 \text{ ft}^3$$

to get the weight of the air inside the ball:

$$w = \rho V = 0.0765 \text{ lb} / \text{ft}^3 \times 33.5 \text{ ft}^3 = 2.56 \text{ lb}$$

The air, inside the ball, is fairly heavy; gravity pulls it down with a force of nearly three pounds! Yet, when you hold a large ball, even with arms outstretched, it does not seem to be heavy. There is another factor to consider—the buoyant force. Objects immersed in a fluid are lifted upward by a buoyant force. The Greek scientist, Archimedes, showed the value of the buoyant force to be equal to the weight of the displaced fluid. When the ball is immersed in air (as it is on earth), it is buoyed up by a force equal to the 2.56 lb. Therefore, we have the nice situation of the downward weight of the ball being 2.56 lb (plus the weight of the plastic shell) and the upward buoyant force on the ball being 2.56 lb. Thus, holding the ball takes very little effort.

However, what is not a 'little effort' is the force required to push the air aside in front of a traveling cyclist.

3.9 Concepts of mass

The word 'mass' is common in our daily vocabulary. You might hear that the mass of the national debt is creating economic problems. Perhaps, in a weather forecast, we hear that a massive snowstorm is imminent. The meteorologist might even say that a cold air mass is moving into the area. Many medical prescriptions are written in doses specified in mass units of grams or milligrams. When terms that are frequently used in everyday speech are encountered in a physics context, we need to be especially careful in their definition and use. The problems with the word mass are compounded further by the fact that the concepts of *mass* and *weight* are unwisely interchanged. To make matters worse, there are actually two kinds of mass in the physical world—*inertial* mass and *gravitational* mass. In this book, when we discuss the mass of an object, it will usually be the inertial mass.

The experiments of Galileo played a critical role in understanding the concepts of mass. Galileo performed a series of very clear-cut experiments involving balls rolling on inclined planes. He observed that the balls speeded up when they were rolling downhill and the balls slowed down as they rolled uphill. Galileo had the insight to conclude that, if the balls were rolling on a level plane, they would continue the motion forever (figure 3.4). Of course, in the real world there are factors (such as friction) that will ultimately end the motion of the ball. Galileo's genius was his ability to understand that the factors, that end the ball's motion, should be considered as separate from the ball's tendency to continue its motion. This tendency for a moving object to remain in motion is called *inertia*.

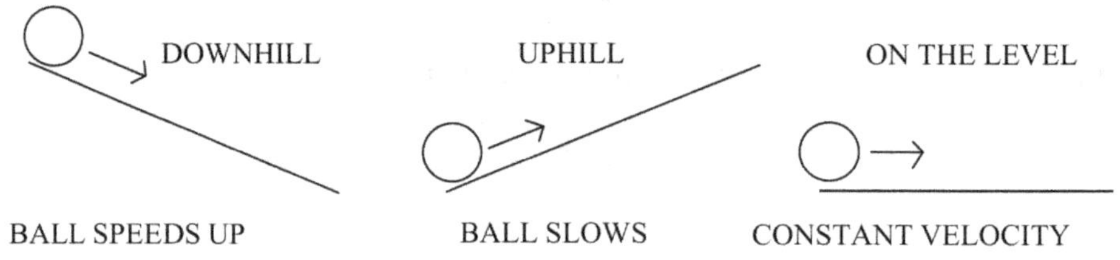

Figure 3.4. Rolling balls.

> This section offers an intuitive description of the physical concept of mass. Inertia is not hard to visualize. Consider two boxes on a table; it is best if they are on well-lubricated wheels. One of the boxes is filled with Styrofoam peanuts and the other is filled with lead bricks. Without lifting the boxes off the table (or tapping on the side), how might you tell which box contains the bricks? Just give the boxes a shove. You will quickly sense that one box offers very little resistance and begins to move easily. The other box requires a substantial effort to set into motion. You could also try to slide the boxes quickly from side to side. In chapter 4, we will call these changes in motion *acceleration*. The resistance you feel in starting, stopping, and changing the direction of the motion is called *inertia*.

This is actually the definition of inertia:

inertia is resistance to a change in motion

The box filled with Styrofoam possesses little inertia; the box of bricks has lots of inertia. We are suggesting that you consider inertia as the intuitive physical concept, and mass as the quantitative measurement of an object's inertia

mass is the measure of inertia

In U.S. Customary units, the mass of an object is measured in slugs; in the SI metric system, mass is in kilograms.

A fair question might be 'why not just lift the boxes in order to determine which contains the bricks?' The answer is that, when you lift the boxes, you are sensing their weight rather than their mass. Weight is a different physical quantity than mass. Weight is the downward pull of the earth's gravity. It is not difficult to conceive a situation in which you cannot sense the weight of the boxes—for instance, an astronaut in orbit would have no sensation of the box's weight. Or ... if you were on a falling elevator, there would be no sensation of weight. However, in both circumstances (the astronaut and the falling elevator) there will still be inertia. The brick-filled box will be just as hard to shake.

The concept of mass is complicated further by the fact that there are two types of mass. *Inertial mass*—is the resistance to acceleration; inertial mass is considered further in Newton's Three Laws of Motion. A second type of mass is *gravitational mass*—the physical quantity that determines the force of gravity. An object's gravitational mass ultimately determines the weight of the object. In chapter 6 (Gravity), we will see a direct connection between the *mass* and the *weight* of a body: a body of mass one slug has a weight of 32.2 lb on the surface of the earth. As an expression:

$$\text{weight}_{\text{in pounds}} = \text{mass}_{\text{in slugs}} \times 32.2 \tag{3.4}$$

Although there is a direct relation between mass and weight, they are different physical phenomena. For instance in cycling, the *mass* of the bike/rider system is of consequence when changing speed (acceleration or deceleration). The *weight* of the system is significant only when cycling on hills.

3.10 Center of mass

With a large or extended object, it is useful to consider that all of the mass of the object is concentrated in one small point—this point is called the *center of mass* (COM). If you throw a small round ball across the yard, its path will sweep out an arc called a parabola. The ball usually spins about its center; however, with a uniform, spherically symmetric object you will probably not even notice the spinning motion. What happens if you throw an object with an asymmetrical shape such as a hammer? Take a hammer and give it a toss across the yard. You will notice a spinning rotational motion of the handle about the hammer's head. While the motion looks complex, the center of mass of the hammer will trace out an arc as smooth as that of the ball. If you watch the motion of the hammer in slow motion, you will see that there is a point, likely a few inches down from the head, which executes the smooth arc. The rest of the hammer twists and turns about this center of mass point.

Many times the center of a mass of an object will be obvious; for instance, the center of a yardstick will be at the 18″ mark. Sometimes, the center of mass will be outside of the body. Consider a doughnut; the center of mass is in the hole. For objects that are flexible, such as a section of a rope, the center of mass will vary depending on the shape. If the rope is held straight like the yardstick, the center of mass is at midpoint; if the rope is bent into a circle, its center of mass is like the doughnut's. Bend the rope into a horseshoe and the center of mass will be near the top of the curve. For a person standing straight up, the center of mass is somewhere in the belly region, bent over and the person is like a horseshoe. In a system of several objects, such as a rider on a bicycle, the center of mass of the system will be determined by the relative masses of the individual pieces. For an adult rider on a modern lightweight bike, the center of mass will be dominated by the configuration of the rider's body.

The center of mass of an object is important because we often consider the motion of an extended body, i.e. a bike rider system, as the motion of its center of mass. We say that the body is treated as a 'point mass'—as if all of a body's mass were concentrated at this single point. When there is rotation of this body, i.e. the rider goes over the handlebars, the rotation occurs about the center of mass. When thinking about the force of gravity on a body, the term 'center of gravity' is appropriate.

3.11 Our standard rider

These values are not represented as any type of average; rather, they are the author's estimate of a compromise between male and female, tall and short, thin and stout, young and not so young, upright and crouched stance, and expensive and inexpensive bikes.

The reader is invited to work all examples with her own personal parameters. The frontal area may be estimated by making a tracing of body outline on kraft paper. When the tracing is cut out—it looks like a large gingerbread cookie—it can be weighed on a kitchen scale and compared to a paper rectangle of known dimensions (two foot by three foot works well). A simple ratio and proportion yields the rider's area. Other more sophisticated methods may be used with digital images of the rider on her bike; CAD and photo manipulation programs allow for determination of body shape geometry.

Of course, there is no standard rider; it might be impossible to find even two riders exactly matched in physical size and cycle. Nevertheless, for purposes of comparison between sections and chapters, most examples in this book are based on a bike/rider system of consistent *standard physical parameters*:

rider weight = 161 lb, rider mass = 5.0 slugs
bicycle weight = 20 lb, bicycle mass = 0.62 slugs
total system weight = 181 lb, system mass = 5.62 slugs
frontal area = 5.38 ft^2

Chapter 4

Linear motion

This chapter begins an exploration of the basic physical concepts regarding the motion of an object. In this chapter, we consider linear or straight line kinematics; in a later chapter, we look at the rotational kinematics of an object moving in a circle. Although these fundamental and instinctive ideas of motion are used on a daily basis, it is important to be precise in their definition and application. Special care is necessary with the terms *speed* and *velocity*. Additional distinction needs to be made between averages of speed and velocity and their instantaneous counterparts.

4.1 Kinematics—the study of motion

Begin by considering a simple journey—suppose we go on a long bike ride, a distance of 50 miles in five hours. The concept of *speed* is defined as:

$$\text{speed} = \frac{\text{distance traveled}}{\text{time}} = \frac{50 \text{ mi}}{5 \text{ hr}} = 10 \text{ mph}$$

Speed is a fundamental factor in describing the trip. With letters, the above expression is written:

$$v = \frac{d}{t} \tag{4.1}$$

with the understanding: v is speed, d is distance traveled, t is the elapsed time

The spirit of the time shall teach me speed.

King John, iv,2 William Shakespeare

doi:10.1088/978-1-6817-4441-4ch4

There are times when the equations might start to look a bit overwhelming. Some folks say they have 'trouble' with the math. Keep in mind that the equations, 'the math', are nothing but a shorthand way of expressing simple phrases and ideas. Now for some subtleties:

4.1.1 Instantaneous speed

If you make this bike ride, the speed, 10 mph, represents your average. It is very unlikely you got on the bike, pressed on the pedal, and traveled at a constant, uniform 10 mph the entire trip. You certainly sped up (accelerated), slowed down (decelerated), stopped at traffic signals, water breaks and so on. Another physical quantity—the *instantaneous speed*—is the speed of the moving object at any given instant (excuse the circular definition). A fair, but not exact, estimate of the instantaneous speed is the reading of the bike's speedometer. While there is still some averaging taking place over a few seconds, the speedometer gives you a feel for the concept of instantaneous speed.

4.1.2 Velocity

Chapter 3 discussed vectors and scalar quantities. It is important to emphasize the distinction between speed and velocity. Speed is a scalar quantity; it makes no reference to a direction. Velocity is a vector; it includes the speed and a specific direction. We often use an arrow to represent the velocity:

$$v = \frac{d}{t}$$

with the understanding: v is velocity, d is displacement, and t is the elapsed time.

Vector quantities with directions are important. Surely, heading '10 mph North' represents a different trip than heading '10 mph South'.

4.1.3 Acceleration

While driving in a car you occasionally hear the phrase 'step on the accelerator'—meaning the gas pedal. We expect the car to have a change in its speed—actually, we should use the term velocity.

As a definition:

acceleration = change in velocity divided by the time for the change.

To write as an equation:

$$a = \frac{\text{change in } v}{\text{change in } t} = \frac{\Delta v}{\Delta t} \tag{4.2}$$

The Greek letter Δ (delta) means change in a quantity

The definition of acceleration is based on velocity—not speed. Acceleration will occur when a velocity vector changes in one of three ways:

- in magnitude—for instance, the car goes from 50 mph to 60 mph. The acceleration is due to change in vector magnitude. This, of course, is a change in speed.
- in direction—a car on a highway might change direction from North to East while maintaining a uniform 50 mph. The acceleration is caused by a change in vector direction.
- in both magnitude and direction. While making a turn on a city street, both the magnitude and direction of the velocity vector probably change.

It is important not to forget the vector nature of acceleration. However, there is still insight to be gained by discussing scalar quantities. Suppose, after stepping on the gas, the car's speed changes from 40 mph to 60 mph in a matter of 4 s. The acceleration is:

$$a = \frac{\text{change in } v}{\text{change in } t} = \frac{20 \text{ mph}}{4\,\text{s}} = \frac{5\,\text{mph}}{\text{s}}$$

The meaning of this 'acceleration' is that for each second of stepping on the gas, the car's speed picks up 5 miles per hour. After 1 second of acceleration you will be moving 45 mph; after 2 s your speed is 50 mph, etc.

Strictly speaking, we are assuming a constant, average acceleration. In chapter 5, when we discuss forces, we will see that there is a major force opposing the motion of the car, namely *air resistance*. Try holding your hand outside an open car window; as the speed increases, you will notice the force on your hand also increases. The consequence of the larger force of air resistance is that it gets harder and harder for the car to gain speed. When the gas pedal is first depressed, the acceleration might be 10 mph/s; but, as the car approaches 60 mph, the acceleration might be only 2 mph/s.

Normally, when the physical laws are first examined, motion is considered in the absence of resistive forces such as air resistance and friction. No doubt, ignoring air resistance in discussing bicycle motion is too idealistic; as soon as you ride faster than a few miles per hour, air resistance is literally in your face.

Another example—how about a car that goes from zero to 60 mph (88 ft/s) in 2.73 seconds? For the acceleration, we obtain:

$$a = \frac{\Delta v}{\Delta t} = \frac{88\,\text{ft/s}}{2.73\,\text{s}} = 32.2\frac{\text{ft}}{\text{s}^2}$$

Wow! What would be the cost of this car? Actually, any car can have this acceleration, even a scrap car without an engine, just push it off the edge of a cliff and it falls under the influence of gravity! We have a special name for the acceleration of gravity—we call it 'g'. The car has an acceleration of one g.

Before leaving this introduction to velocity and acceleration, consider the motion of a bicycle. Suppose a very strong cyclist is able to go from rest to 20 mph (29.3 ft/s) in 10 s. The acceleration is:

$$a = \frac{\Delta v}{\Delta t} = \frac{29.33\ \text{ft/s}}{10\ \text{s}} = 2.93\ \text{ft/s}^2 = \frac{2.93\ \text{ft/s}^2}{32.2\ \text{ft/s}^2} = 0.091\ \text{g}$$

Although the expression we have been using for the acceleration is fine, occasionally, we will need a bit more sophistication. Instead of using Δv, consider a notation in which we go from some initial speed v_i to a final speed v_f. The final speed is the sum of the initial speed and the contribution from the acceleration.

$$v_f = v_i + a\,t \tag{4.3}$$

Consider again our rider who was capable of the acceleration 2.93 ft/s^2. Suppose she was already traveling at 15 mph (22 ft/s) when she kicked in with the ten-second acceleration. What is her final speed at the end of the acceleration period?

$$v_f = v_i + a\,t = 22\ \text{ft/s} + (2.93\ \text{ft/s}^2 \times 10\,\text{s}) = 22\ \text{ft/s} + 29.3\ \text{ft/s} = 51.3\ \text{ft/s}\,(35\,\text{mph})$$

An acceleration from 15 mph to 35 mph will encounter much greater air resistance than the same acceleration from zero to 20 mph. It would take superwoman to maintain such an acceleration. Let us push the ideas a bit further.

In the case of going from 15 mph to 35 mph, is there a valid meaning to an average speed? Yes, as long as the acceleration is constant—a steady increase in speed. We can take the starting value v_i and the ending value v_f and divide by two. The average speed v_{ave} during this period is:

$$v_{ave} = \frac{v_f + v_i}{2} = \frac{35\,\text{mph} + 15\,\text{mph}}{2} = 25\,\text{mph}$$

We can obtain a more general expression by noting:

$$d = v_{ave} \times t$$
$$d = \frac{v_f + v_i}{2} \times t = \frac{(v_i + at) + v_i}{2} \times t \tag{4.4}$$

Grouping terms:

$$d = v_i t + \frac{1}{2}\,a\,t^2 \tag{4.5}$$

The result is a nice, simple expression that allows us to determine how far we have traveled while going from rest ($v_i = 0$) to 20 mph (29.3 ft/s):

$$d = \frac{1}{2}at^2 = \frac{1}{2} \times 2.93\ \text{ft/s}^2 \times (10\ \text{s})^2 = 147\ \text{ft}$$

What would be the result using the average speed of 10 mph (14.7 ft/s)?

$$d = v_{\text{ave}}\, t = 14.7\ \text{ft/s} \times 10\ \text{s} = 147\ \text{ft}$$

The results agree! So far, our examples have been situations in which the speed increases (accelerations); of course, moving objects eventually have to slow down (decelerations). All of our previous definitions and equations in this chapter apply to both acceleration and deceleration.

In the case of objects speeding up, the acceleration is *positive.* When objects slow down, the acceleration is *negative.* Suppose our cyclist traveling at 20 mph (29.3 ft/s) comes to a stop in 10 s, the deceleration is:

$$a = \frac{\Delta v}{\Delta t} = \frac{v_\text{f} - v_\text{i}}{t} = \frac{0 - 29.33\ \text{ft/s}}{10\ \text{s}} = -2.93\ \text{ft/s}^2$$

4.2 Headwinds and tailwinds

Do head winds and tail winds cancel in a round trip?

Consider a rider with who sets out on a long trip and encounters a head wind that cuts her speed to 15 mph on the outward leg. On the return trip, the wind is a tailwind that results in a speed of 25 mph. Will her average be?

$$\frac{15\ \text{mph} + 25\ \text{mph}}{2} = \frac{40\ \text{mph}}{2} = 20\ \text{mph}$$

The answer is no! A simple average is not correct; she does not spend an equal time on the outbound leg versus the homeward leg. The return trip, at the higher speed, will be a shorter time.

Think of the following analogy—in a badly designed grading scheme, a course has only two exams—a mid-term, worth one fourth of the final grade, and a final exam worth the remaining three fourths. A student attains a perfect 100% on the midterm, gets overconfident, and stops studying. He then proceeds to get 50% on the final exam. What is his course grade? Will it be the simple average of the two exams 75% for a C+ in the course? Unfortunately, no, a *weighted average* needs to be calculated. The final exam is given more weight. The 100% on the midterm earns the student 100 points; the 50% on the final earns the student only 150 points. Totaling his points, we get 250 points out of the maximum of 400. His final grade percentage is just 62.5%. At best, the student receives a D in the course.

How do we create a weighted average for our cyclist with head winds and tail winds? Let us put some distances on this journey. Assume she rides 40 miles out at15 mph, turns around, and rides 40 miles back home at 25 mph.

The longest road out is the shortest road home.

Irish Proverb

Obtain the travel times, for the outward leg of the trip: $v_1 = 15$ mph

$$t_1 = \frac{d}{v_1} = \frac{40 \text{ mi}}{15 \text{ mph}} = 2.67 \text{ hr}$$

for the homeward leg: $v_2 = 25$ mph

$$t_2 = \frac{40 \text{ mi}}{25 \text{ mph}} = 1.60 \text{ hr}$$

The total time is: 2.67 hr + 1.60 hr = 4.27 hr
The average speed is:

$$\text{average speed} = \frac{80 \text{ miles}}{4.27 \text{ hr}} = 18.7 \text{ mph}$$

The average is less than the simple average of 20 mph. The outward leg, at the slower speed, is weighted more heavily than the return trip speed. The effect is even more dramatic if a very strong wind results in a larger speed discrepancy. A wind that results in a speed 5 mph on the way out and 35 mph on the way home will produce an average speed of only 8.75 mph. The simple average of 20 mph is even more inaccurate!

4.3 Riding uphill and downhill

How about riding uphill and then riding downhill? Do the hills cancel in a roundtrip?

The ride downhill is usually a lot shorter time than the ride uphill; the uphill and downhill segments do not cancel. The analysis is similar to the head wind and tail wind situation. The times of the uphill and downhill legs will not be the same (assuming equal efforts by rider). Going fast downhill in a short time interval does not cancel the long, slow uphill climb.

Chapter 5

Forces—Newton's laws of motion

In this chapter, we take major steps toward a discovery and appreciation of the fundamental laws of nature that underlie so much of the bicycle's magical behavior. The primary emphasis is to understand the role of forces in controlling the motion of the two-wheeler. A rationale is developed for the mechanism by which the bicycle offers the most efficient form of human powered locomotion. The force of friction, normally considered a foe to motion, is regarded as a friend. We will learn how friction accelerates the bicycle forward and how this same friction stops the bike when needed. Freewheeling, the ability of the bicycle to joyously coast along with no effort by the rider, is a benefit of inertia—the tendency of an object to remain in motion.

Every rider knows there is effort needed to propel the machine into motion and to maintain its motion against major resistive forces such as air resistance and gravity. Our rider must also counter the less significant forces such as the rolling resistance of the tires and the bearing resistance of the hubs. This chapter reveals the manner by which these forces develop and evaluates their relative significance.

The operation of the bicycle is shrouded in mystery. The bike, inherently unstable with a large mass perched above tiny points of support, develops an amazing stability once set into motion. Most fascinating is that this stability exists with or without input from a rider. The stability is a consequence of nature's basic laws of motion.

In chapter 5, we explore Sir Isaac Newton's three laws of motion. These laws, that establish the relationship between forces and motion along with the principle of universal gravitation, were promulgated in 1687 by Newton in his *Principia Mathematica*. The *Principia* is arguably the most important scientific text ever written. Newton's brilliant mind brought a concise, comprehensive, and beautifully simple approach to understanding motion. The science that studies motion and its causes is called dynamics. Newton's laws are the foundation of dynamics. These laws are very easy to describe, nevertheless, extremely profound and universally significant. Newton's laws are used to calculate the motion of a falling apple or the

doi:10.1088/978-1-6817-4441-4ch5

orbits of the planets. They were also the basis in determining the trajectories of space flights that landed astronauts on the moon and, perhaps more importantly, got them safely back to earth. These same Newton's laws of motion govern the astonishing behavior of the bicycle as the rider effortlessly balances on two wheels and glides with minimal exertion.

Difficulty sometimes arises in applying Newton's laws because we forget how straightforward they really are. Occasionally, you might want to reread these laws and give their simple beauty a chance to be slowly absorbed. The laws of motion are not highly mathematical; in fact, they may well be the mathematically simplest of all physics. Two of the three laws do not even involve an equation. One of the nicest aspects of Newton's laws is that they can be applied to understand events in our daily experience. It is easy to encounter and experiment with the laws at will; no expensive laboratory equipment is needed. A bicycle, with an economical speedometer–odometer, is a traveling physics laboratory.

As the motion of a bicycle is examined, we will learn that there are numerous forces affecting the two-wheeled contraption. These forces are significant in both the acceleration and the deceleration of the system. Some of these forces result from the bike/rider interaction with the environment. External forces occur as a consequence of the tire's contact with the ground and some arise from the necessity of the bicycle/rider system to move air aside. Gravity also has a major effect as the bicycle travels up and down hills. Some forces are associated with the deformation of a soft tire; these are described as rolling resistance. There are internal forces, for instance, the rider pressing on the pedals or the bearings rubbing in hubs. Some forces get a special name such as the centripetal forces that cause the bicycle to make a turn. Frictional forces describe various types of tire/wheel/road contact forces. The complexity of the forces on the bicycle, while not difficult to understand, is subtle and requires careful thinking and precise choice of words.

When considering the forces on a bike/rider system, it is important to clearly define which bodies are part of the 'system'. Forces that act between components of the system are called *internal forces*. Forces that arise from interactions outside of the system are called *external forces*. For example, if we define the system to be the bike and the rider, then the thrusts of the rider's feet on the pedals are internal forces—within the system. In addition, the pressure of the brake pads on the wheel rims is considered an internal force. The external forces on the system are forces from the road, gravity, air, etc. According to Newton's First Law, the motion of the bicycle can only be changed through the action of external forces. Before studying the laws of motion, it is recommended that the reader review the concepts of inertia and mass in section 3.9.

In chapter 2, we defined a *force as a push or pull*. We also saw that vectors are a very useful way to understand forces. The image of a person pushing or pulling on a table is easy to visualize. Newton's three laws enable us to understand the manner in which forces effect and affect the motion of an object. As we consider Newton's laws, we will see their application to the table and, for more fun, the bicycle as it moves in all sorts of circumstances: slow and fast, uphill and downhill, with the wind and against the wind.

5.1 Newton's First Law of Motion

The First Law of Motion is stated as:

A body at rest will remain at rest, or a body in motion at a constant velocity will remain at a constant velocity unless acted upon by a net external force.

Sounds simple enough? It is—no equations, no calculations! We have actually seen a version of this law in chapter 3. The First Law of Motion is a restatement of Galileo's concept of inertia, the tendency of a body to maintain its constant velocity. Recall that *mass is a measure of inertia*. Although the First Law of Motion does not use an equation, it is loaded with significance and subtlety. Let us dissect a few of its profound implications:

1. A body at rest or moving at a constant velocity is *not* accelerating. Whether the bike is parked in a garage or traveling down a straight highway at 25 mph, the acceleration of the bike is zero. An alternative way to state the First Law is:

   An acceleration of a body requires a net external force.

2. The First Law uses the word 'velocity'—recall from chapters 3 and 4 there is a very significant *distinction* between speed and velocity. An object's speed might be stated as '20 mph'. The concept of velocity is a vector that combines the speed with a direction. A velocity vector has both magnitude (20 mph) and direction (e.g. West).

   When it comes to vectors, the direction is of equal importance as the magnitude.

   The implication of the above is that it takes a *net external force to change a body's direction of motion.* If you ride a bike and make a turn to the left, it requires a net force to the left. Sometimes, cyclists will say that they turn by leaning. Leaning is not the force to effect a turn. While the basics of forces are examined in this chapter, a complete analysis of turning is in chapter 12.

3. It takes a *net external force* to change an object's motion. The term *net force* is based on the concept of vector addition (chapter 2). If you have two people on opposite sides of the kitchen table each pushing with 50 lb of force, the net force is zero. If there is no net force, a table at rest will remain at rest.

   Often, we refer to the net force as the 'resultant force'—sometimes, it is simply called the resultant.

4. The use of the term *external force* is also a critical aspect of the First Law of Motion. When applying Newton's laws, we invoke the concept of 'the system'. The system is a collection of objects (or bodies) considered to be associated with each other. Once we define the objects that comprise the system, we then decide which forces are 'external' to the system and which forces are 'internal' to the system.

   In the application of Newton's laws, a decision must be made that clearly defines the specific bodies that comprise the 'system'. Forces that act on the system from outside are considered 'external' forces. It requires a net external force to cause an acceleration of the system.

With a system definition of 'the kitchen table' and the people as 'external forces'—if one person pushes with a force of 50 lb and the other pushes with a force of 20 lb, there will be a net external force on the table of 30 lb. Newton's First Law states that the table will not remain at rest because of the net external force. The table will begin to accelerate; to describe fully the motion we need Newton's next law of motion.

The amazing thing is that Newton's laws apply regardless of the bodies chosen to be included in the system. We simply have to be consistent in deciding internal versus external forces. In most cases, the objects to be included in the system are intuitively obvious. If we wish to study the motion of the table, the system is the table. If we wish to study the motion of a cyclist, the system is the rider plus the bicycle. Sounds easy? Maybe. If the system is the rider plus bike, we must acknowledge that the effort of the rider pushing on the pedals is an internal force—not an external force! A force on the pedal is not the force of the bike's acceleration. Nor is the brake pad rubbing on the rim the force of deceleration. What then causes the bicycle to start and stop? You will have to keep reading; there is much subtlety involved.

With regard to the kitchen table—the external forces would be due to the people pushing, the force of gravity, an upward push by the ground, and likely a force of dragging called friction. Conceivably, if the table were moving fast enough, we would need to consider a force of air resistance. The primary external forces on the bike are gravity, ground, friction, rolling resistance, and air resistance. We will explore each of these in detail.

Where might we best experience Newton's First Law of Motion on a bicycle? Certainly, the most fun instance is cruising down a moderate hill at a constant speed. The downward pull of gravity offsets the upward resistive forces, and the rider is able to 'fly' happily down the hill.

5.2 Newton's Second Law of Motion

Newton's Second Law of Motion establishes the relationship between the *net external force* on a body and the *acceleration* of the body. As with the First Law, the Second Law is simple and concise. It is stated as:

The acceleration of a body is directly proportional to the net external force acting on the body and inversely proportional to the mass of the body.

The First Law tells us that a net external force causes accelerations. The Second Law allows for the determination of the magnitude of the acceleration. If the net external force is doubled, the acceleration is doubled. Newton's Second Law also states that the acceleration is inversely proportional to the *mass*. This means, if you double the mass, the acceleration will be halved.

A simple written expression of Newton's Second Law is:

$$a = \frac{F_{\text{net}}}{m}$$

where a is the acceleration—the change in velocity with time, F_{net} is the net external force—the vector sum of all forces external to the system and m is mass—the measure of inertia (the resistance to change in motion).

The expression, simple as its looks, has a very important subtlety. Both force and acceleration are vectors. Careful consideration reveals that what appears to be one equation is actually three equations. Forces are vectors, maybe three-dimensional vectors, and acceleration can be a three dimensional vector. There can be an independent Second Law for each of the three directions (dimensions)—x, y, z (left and right, up and down, back and forth). Vertical forces determine vertical accelerations; horizontal forces determine horizontal accelerations.

Fortunately, most of a bicycle's motion can be described in terms of one (when traveling in a straight line) or two (when making a turn or riding a hill) directions. Also helpful is that the three directions are perpendicular to each other. Thus, forces and motion in the horizontal do not affect forces and motion in the vertical. Forces and motion in the left and right direction do not affect forces and motion in the forward direction. Each vector's component is independent of the others. The reader might wish to review the discussion of vector components in chapter 3.

These ideas are so important that we will restate them in slightly different words:

- The horizontal acceleration is only determined by the net horizontal force. The vertical acceleration is only determined by the net vertical force.
- An acceleration to the left requires a net force to the left; an acceleration to the right requires a net force to the right.

Can I turn a bike just by leaning?

The answer is no. Later, when we explore the turning behavior of the bicycle, it will be useful to review the above statements. We will see that a bike making a left hand turn is accelerating to the left. Newton's Second Law states that an acceleration to the left requires a force to the left. While the lean of a bicycle is important during a turn, it does not create the leftward force. The force on your leaning body is still vertically downward—the downward force of gravity, aka your weight.

5.3 Units of force, motion, mass

Before moving on to the next Law of Motion, it is best to deal with issues of units. If you sometimes get frustrated with issues of units, please review the discussion on units in chapter 3. In the *United States Customary* units, the following fundamental units are used: force is specified in pounds (lb); acceleration is specified in feet per second squared (ft/s^2); mass is specified in slugs. Thus, a force of one pound acting on a mass of one slug produces an acceleration of one ft/s^2.

In the next chapter, we learn that, on the surface of the earth, the acceleration of gravity is 32.2 ft/s^2; one slug of mass weighs 32.2 pounds. Suppose we had a cyclist + bike system of combined weight w = 181 lb. In order to apply Newton's Second Law, we must get her mass in slugs:

$$m = \frac{w}{32.2} = \frac{181 \text{ lb}}{32.2 \text{ ft/s}^2} = 5.62 \text{ slugs}$$

If she experiences a net forward force of 10 lb, Newton's Second Law tells us that her acceleration will be:

$$a = \frac{F_{\text{net}}}{m} = \frac{10 \text{ lb}}{5.62 \text{ slugs}} = 1.78 \text{ ft/s}^2$$

Using concepts of motion from chapter 4, if she starts from rest and the force of acceleration remains constant for 15 s, her speed will be:

$$v = a\,t = 1.78 \text{ ft/s}^2 \times 15 \text{ s} = 26.7 \text{ ft/s} = 18.2 \text{ mph}$$

Incidentally, our rider must be quite strong; later in this chapter, we will see that an opposing force of air resistance begins to build rapidly after about five mph. Maintaining the constant forward force becomes increasingly difficult and ultimately impossible.

5.4 Newton's Third Law of Motion

This section explores the last of Newton's laws of motion. As with the first two laws, a statement of the Third Law is not difficult. Of the three laws, the Third Law of Motion is the most commonly quoted in everyday conversations. A simple, precise statement of Newton's Third Law of Motion is:

If body A exerts a force (the action) on a second body, B, then body B exerts an equal and opposite force (the reaction) on body A.

> Important points to consider: forces always occur in pairs between two bodies; the forces are equal in magnitude; the forces act on different bodies. One force acts on body A; the other force acts on body B. *Opposite force* means opposite in *direction*. If body A pushes to the right, body B pushes back toward the left. The second force, the so-called reaction force, is a real force.

Is the force actually equal and opposite in all cases? Yes! Even if you wad up a small piece of paper and exert a force of 10 lb in throwing it across the room, the paper will exert an equal and opposite force on you of 10 lb. The paper might only

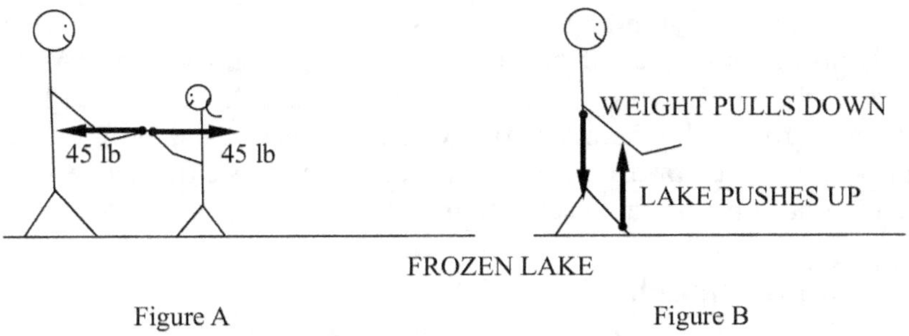

Figure 5.1. (A,B) Father and daughter on ice.

weigh a few ounces, but it will push back on you with whatever force you exert on it! To gain further insight in the Third Law of Motion, let us examine a situation of two ice skaters (figure 5.1A), a 161 lb man (mass of 5 slugs), and his daughter who weighs 40.25 lb (mass of 1.25 slugs).

Suppose they stand facing each other, initially close together. With their palms touching, one or both push off with horizontal forces. It does not matter who is the stronger or who does the actual pushing. If he pushes on her with a force of 45 lb, by Newton's Third Law she will exert an equal and opposite force on him of 45 lb. Regardless of how the pushing is done, the force on each skater will be equal and opposite. Since they are on ice, there are minimal frictional forces that can be ignored.

What can be said about the motion of the skaters? Since the forces are horizontal, we calculate their horizontal accelerations using Newton's Second Law:

$$a_{\text{man}} = \frac{F}{m} = \frac{45 \text{ lb}}{5 \text{ slugs}} = 9 \text{ ft/s}^2 \qquad a_{\text{girl}} = \frac{F}{m} = \frac{45 \text{ lb}}{1.25 \text{ slugs}} = 36 \text{ ft/s}^2$$

The man has four times the mass of the girl and experiences the same force as the girl; his acceleration is one quarter of the girl's acceleration. How long are these accelerations maintained? From Newton's First Law, we conclude that the man and his daughter will only accelerate while they experience the horizontal forces. They accelerate until their hands separate and, in the absence of resistive forces such as friction and air resistance, they will maintain speed until reaching shore. How fast will they be moving at the time of separation? Suppose they exert forces on one another for 0.5 s. From chapter 4 we obtain an expression for their final speeds at the time of separation.

$$v_{\text{man}} = a_{\text{man}}\ t = 9 \text{ ft/s}^2 \times 0.5 \text{ s} = 4.5 \text{ ft/s}$$
$$v_{\text{girl}} = a_{\text{girl}}\ t = 36 \text{ ft/s}^2 \times 0.5 \text{ s} = 18.0 \text{ ft/s}$$

The girl ends up traveling four times faster than her father.

Until now, we have only looked at the horizontal forces and motion. Can we use Newton's laws to understand the vertical vectors? Yes! Newton's laws work in all directions. Consider the father whose weight is 161 lb (figure 5.1B). This means that the earth pulls downward on the man with a force of gravity of 161 lb. Why does this force not produce a vertical acceleration? It would, if the man were in mid-air with no upward force to resist the pull of gravity. The result is called falling down!

However, the man was standing on ice—as long as the ice is nice and thick, the outcome is no different than if he were standing on solid ground. As he pushes downward on the ice, Newton's Third Law tells us that the ice will push equal and opposite—upward, on the man. The two vertical forces on the man, the downward pull of gravity and the upward push of the ice, cancel. Since the net vertical force is zero, the man has no vertical acceleration.

5.5 Role of arm muscles

Do the arms and upper body play any role in delivering force that contributes to the forward motion of the bike?

During the evolution of the bicycle in the mid-19th century, attempts were made to add an arm-powered driving mechanism to the 'Hobby Horse' (the Draisine). A simple observation as to the relative size of the arm muscles versus the leg muscles shows the inferiority of an arm propelled machine. In addition, the upper half of the body is occupied with steering and, in the case of hand brakes, stopping the bicycle. Nevertheless, in a modern bicycle where the major effort is delivered by the lower body, the arms do add some force toward the bike's propulsion. Once again, Newton's laws are the key to understanding the force contributions of the arms. Consider that the rider pulls upwards on the handlebars when pedaling very hard. From Newton's Third Law, the handlebars push down on the hands—this downward push by the handlebar adds to the force that can be applied to the pedal.

A free body diagram of a rider standing and putting all of his weight onto one pedal and simultaneously pulling up on the handlebars is shown in figure 5.2. The image on the left includes the rider and the handlebars and pedal. Figure 5.2B shows only the forces on the rider.

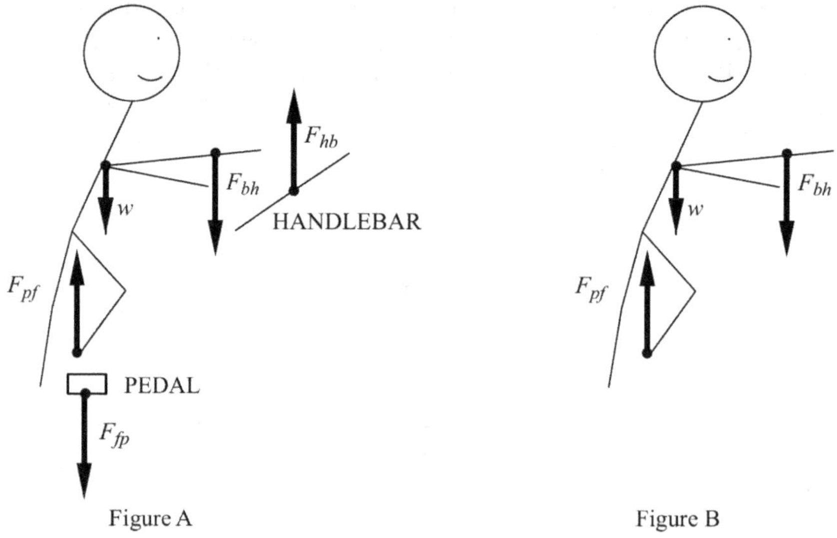

Figure 5.2. (A,B) Man forces.

There are two action–reaction pairs as expected from Newton's Third Law.

At the handlebar: F_{hb}—the hands pull up on the handlebar and—the handlebar pushes down on the hands F_{bh}.

At the pedal: F_{fp}—the foot pushes down on the pedal and—the pedal pushes up on the foot F_{pf}.

Gravity pulls down on the rider with a force equal to his weight w. Newton's Second Law tells us that if there is no vertical acceleration, the vertical forces on the rider must add up to zero:

$$-w + F_{pf} - F_{bh} = 0 \qquad \text{or upon rearranging} \qquad F_{pf} = w + F_{bh}$$

If the rider puts all of his weight of 161 lb on one pedal and pulls up on the handlebar with a force of F_{hb}= 50 lb (thus F_{bh} = 50 lb), the pedal must push up on his foot with a force F_{pf} of 161 lb+50 lb = 211 lb. From Newton's Third Law, he then pushes down on the pedal with a force of F_{fp} = 211 lb.

The force with which the rider pulls upward on the handlebar is effectively added to his weight. The consequences of applying such a large force to the pedal while pulling upward on the handlebars is examined later when we look at the gearing system in chapter 11.

5.6 Frictional forces—a simple model

Think of trying to push a heavy box across a rough floor. Start with a gentle push; the box does not budge, perhaps a slightly stronger push—still no movement. Eventually, if you are strong enough and the floor releases its grip on the box, it begins to move. You might notice that the force to keep the box in steady motion is less that the force it took to break it free. Welcome to the phenomena of friction. Friction is an omnipresent force that develops between surfaces. It is difficult to see and loaded with complexities easily misinterpreted. This simple act of pushing on the box needs elaboration. At first, with the easy pushes before the box budges, we say there is *impending* motion of the box. Once the box breaks free, there is *actual* motion. Another aspect of the friction force is that it is always in *opposition* to the direction of impending and actual motion. The definition of friction:

friction is a force between surfaces that opposes motion or impending motion.

For the box along the floor, force is a horizontal vector. If the push is to the right, the force of friction points to the left. A push to the left results in a friction force to the right. The definition of friction states that it opposes motion—sometimes it succeeds in its opposition; sometimes it does not succeed.

When the box has yet to move and the force of friction is opposing impending motion, it is called *static friction*. Once the box is actually moving, the frictional opposition is called *sliding friction*. In general, static friction is greater than sliding friction.

How might friction be visualized? In a simple model, think of the bottom of the box and the floor as having rough surfaces in contact (figure 5.3A). If there is no attempt at horizontal motion, the rough surfaces are just meshed together and there

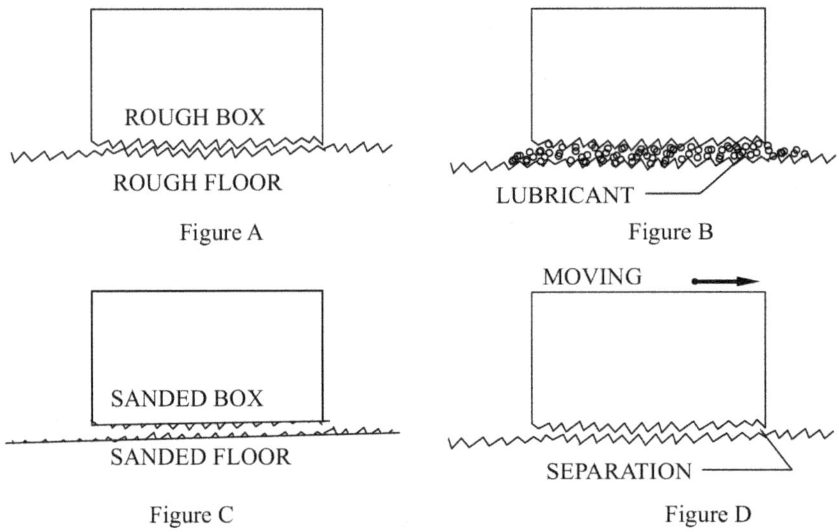

Figure 5.3. (A,B,C,D) Rough box on floor.

is no horizontal force of friction. A moment's reflection of figure 5.3A should assist in understanding how the force of friction only develops as we attempt to slide the surfaces over one another. The friction force arises as the ridges are compelled, by a horizontal shove, to ride over one another.

The presence of a lubricant such as oil, grease, wax, or even water will reduce the force of friction. Have you ever slipped on a wet or recently waxed floor? The lubricant consists of small particles filling in the hills and valleys of the rough surfaces. Obviously, an effective lubricant is one in which the small particles are able to roll and slide past each other (figure 5.3B). Another technique to reduce friction is to lessen the roughness of the contact surfaces (figure 5.3C). Sanding the surfaces lessens the heights of the hills and valleys. The smaller hills and valleys are less able to resist horizontal forces trying to move the box.

> This simple model of friction also helps in understanding the reason *static friction is greater than sliding friction*. More force is required to get the box moving than is needed to maintain steady motion. Static friction is visualized as having the rough surfaces fully meshed together (figure 5.3A). Whereas, once the body is in motion, the rough surfaces 'ride' over one another; they do not get a chance to settle in and mesh (figure 5.3D). Only the tips of the rough edges have an opportunity to catch on each other.

5.7 Static and sliding friction

Think again, what happens when we try to push a heavy 250 lb box across a rough but level floor. If the pushing force is parallel to the ground, it has no vertical component that will add or subtract from the box's weight. In figures 5.4, P is the force pushing on the box; w is the weight of the box; f is the opposing force of friction

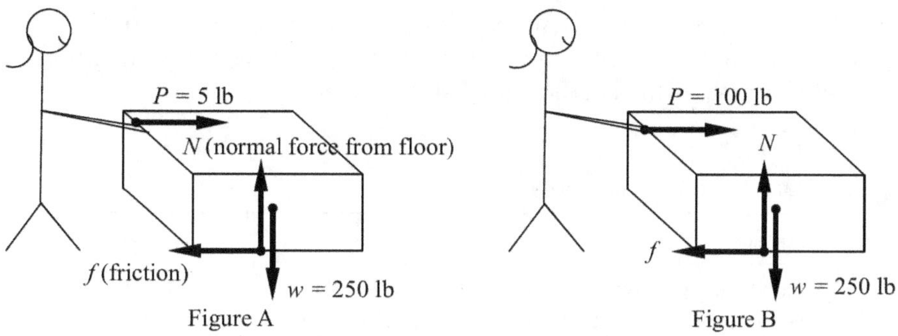

Figure 5.4. (A,B) Pushing a box.

(it is traditional to use a lower case *f* for friction). *N* is the upward (normal) force from the ground. Notice that the direction of friction is opposite to the direction of the pushing force.

From the experimentally observed behavior of friction, the force of friction is modeled as:

- The force of friction does not depend upon the area of contact. Thus, a box of wood sliding along, with its large face in contact with the ground, experiences the same friction as if it had been sliding along a small face. No doubt the broad surface in contact with the floor engages more ridges hooking together; however, the ridges are not pressed as tightly together as when the box is resting on a small face. This counterintuitive concept has application when we think about a comparison in frictional forces between a narrow tire and a wide tire in contact with the road. Water on the road and grooved tires invalidate this simple model.
- The force of sliding friction does not depend on the speed of the surfaces. To the cyclist this is a good thing; it would be an unpleasant surprise if friction vanished on a high-speed cycle!
- The force of friction is directly proportional to the *force pressing* the surfaces together. This is the *normal* force. (*Normal* in this context means perpendicular to the contact surfaces—on level land, a telephone pole is normal to the ground). Thus, we say:

$$f \sim N$$

This expression is made into an equation with the introduction of a proportionality constant μ, the *coefficient of friction.*

$$f = \mu N$$

In order to avoid a common error, an important aspect of static friction is that the above expression represents the maximum value of static friction. If no one is pushing on the box, the force of friction is zero. Until the box breaks free, the force of static friction is only as large as the applied push. The expressions for friction are best stated:

static friction $\quad f_{\max} = \mu_{\text{static}} \times N \quad$ or $\quad f \leqslant \mu_{\text{static}} \times N \quad$ (5.1)

sliding friction $\quad f = \mu_{\text{sliding}} \times N \quad$ (5.2)

Values of the frictional coefficient have been measured for a variety of contact surfaces. A few examples are listed in table 5.1. The coefficient of sliding friction can be as small as one-half the coefficient of static friction.

Table 5.1. Coefficients of friction.

Surfaces	μ_{static}	μ_{sliding}
Rubber on concrete/blacktop	0.9–4 [1]	0.5–0.8[a]
Wood on wood (oak)	0.54[a]	0.32[a]
Ice on ice	0.05–0.15 [1]	0.02 [1]

[a] www/roymech.co.uk/useful_tables

A caution: measurements on the coefficient of friction are dependent on the condition of the samples. Factors such as roughness of wood, presence of liquids, and oxidation of metals will cause variation in the frictional coefficients. The values should be considered as approximate; various references show discrepancies in their listed values [1]. Missing from the table are coefficients for rubber on ice. No doubt important, but this coefficient is greatly affected by variables such as the hardness of the rubber and the condition of the ice; it is difficult to find accepted values; values of 0.1–0.2 are ballpark estimates.

There are several pitfalls when trying to understand a situation involving friction. For example:

- While we say that friction *opposes* motion or impending motion, it can also be the force of acceleration—in other words, it *causes* the motion. For instance, it is incorrect to say that friction from the ground always opposes the motion of the bicycle. We will see in a later section of this chapter that static friction, which arises between the ground and the tire, is the net external force that accelerates the bicycle! Thus, friction does not always slow the bike down—it is also responsible for speeding the bicycle up. In the next few sections, we explore the mechanism whereby friction in opposing motion actually causes motion for walking and cycling.
- Static friction is also the force that is most effective in stopping the bicycle. In a normal stop—one that does not involve any skidding of the tires—it is the force of static friction between the ground and the tires that causes the deceleration of the bicycle.
- Skidding of the tires is undesirable since it involves sliding friction. Examination of table 5.1 reveals that the coefficients of sliding friction are much smaller than the coefficients of static friction. The effectiveness of braking is greatly reduced during a skid. A skidding stop can involve a coefficient of friction as little as one-half of a normal, non-skidding stop. If the friction coefficient is one-half, the frictional force will be one-half and the stopping distance will be twice as long in a skid. There is a very good reason cars are now equipped with anti-lock brakes that activate during a skid.

Does sliding friction play any role in a normal stop? Yes, it is a force of sliding friction that exists between the spinning rims and the brake pads.

- Static friction is a non-dissipative force. It does not generate heat between the contact surfaces. There will be no heating of the bottom of the tire or the ground during a deceleration or acceleration caused by static friction.
- Sliding friction is a dissipative force that does generate heat. The force between the brake pads and the rims is an example of sliding friction. The force causes the rims and brake pads to become noticeably warm to the touch during a hard stop. The heating effect may also be readily observed in disk brakes. Carefully touch the rotors of disk brakes to experience the large amount of heat generated during a stop.

As another example of the difference between static and sliding friction—have you ever come inside on a cold day and tried to warm your hands by rubbing them against one another? As the hands slide back and forth with a force of sliding friction between them, you will notice the heat generated and nice warm hands. In contrast, if you merely press your hands together with insufficient sliding effort, the force of static friction does not warm the hands.

5.8 Friction as the propulsion force in walking

How can a force of friction, usually described as a force that opposes motion, be the cause of a body's forward motion?

I know him by his gait: he is a friend.

Julius Caesar, i,3

Before we try to understand the motion of the bicycle, let us think about the propulsion forces of walking. This will assist in understanding the role played by the various types of friction in the motion of the bicycle.

In order to walk toward the right, the foot must push on the ground toward the left (figure 5.5A). Since our walker is also pushing with his weight onto the floor, the resultant force on the floor is down and to the left F_{fg}. If the floor were covered with marbles—they would be pushed backward. Your foot would also slip backward. Now of course, floors are usually not covered with marbles and your foot does not slip.

A force of static friction resists the impending rearward motion of your foot. This frictional force points in the forward direction! The foot pushes on the ground downward and toward the rear, and from Newton's Third Law, the ground pushes up (normal force N) and in the forward direction (frictional force f) on the foot. It is this force of friction from the ground that causes your forward acceleration. It is not *your* force on the ground that causes the motion; rather, it is the *ground's* force on you!

As you continue to walk, taking step after step, your foot is repeatedly placed in contact with the ground. In normal walking, the frictional forces between the foot and the ground are *static friction*. The ground is stationary, and as the foot pushes on the ground, it is stationary with respect to the ground—otherwise you are slipping! We will eventually see that a moving bicycle is much the same. Although it

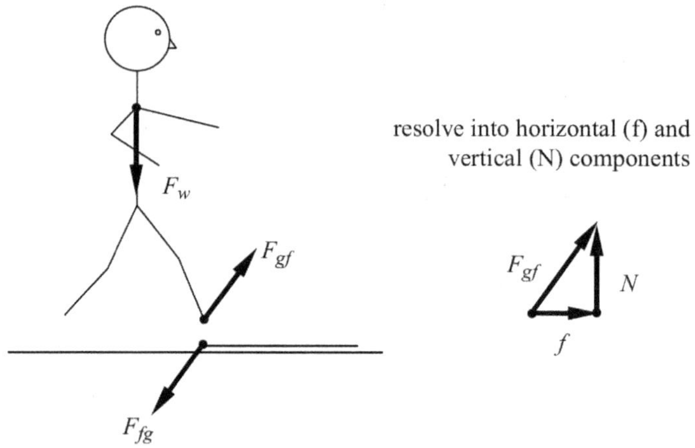

Figure A

Figure B

Figure 5.5. (A,B) Walking man.

is traveling down a road, the bottom of the tire is at rest in contact with the ground. The force of static friction between the road and the rear tire propels the machine.

Walk a few steps and think about the process described above. You might wish to envision the result if the coefficient and force of static friction were too small, a common occurrence on ice.

The process of walking does not consist of a continuous, sustained acceleration across the room. For one thing, the acceleration ceases when the foot breaks contact with the ground. Even more significantly, once we have made all this effort to *accelerate* the foot and attached leg, we must then *decelerate* the leg and bring the foot to rest as it swings forward and strikes the ground. Remember, the ground is at rest and your feet are at rest as they touch the ground. The situation with the feet *not* at rest is called 'slipping'. Our feet do not ordinarily slip as we walk!

These accelerations and decelerations require muscular forces and energy expenditure on the part of the walker. The use of a continuously rotating wheel and a rotating leg powering bicycle pedals is much more efficient since there are no decelerations.

5.9 The acceleration and deceleration of the bicycle

How does the force of pressing on the pedals lead to the acceleration of the bike?

> Newton's First Law of Motion states that it takes a *net external force* to produce a change in a body's velocity. Since we usually treat the rider and bicycle as a single body—'the system'—any force between the rider and the pedals, or the rider and the handlebars, or the rider and the saddle, etc, are considered *internal forces* within the system. Internal forces cannot effect changes in the velocity of a body. The rider's force on the components of the bicycle cannot be the force that produces the system's forward acceleration. To produce acceleration, starting the bike from rest or changing its velocity, a net external force is needed. If the change in motion is *horizontal*, the *net external force needs to be horizontal.*

There must be some relationship between the rider force on the pedal and the net external force as required by Newton's First Law. The explanation of the transmission of the pedaling force to the acceleration force of the bicycle involves a discussion of the cranks, the chain, the gears, the hub, the spokes and, ultimately, the wheel. It is best to consider the transmission components in a later chapter (chapter 11). For now, let us keep things as simple as possible—instead of using the pedals, just reach down and give the rear wheel a spin with the hand.

5.9.1 Acceleration when starting from rest

Consider a rider heading to the right (figure 5.6A).

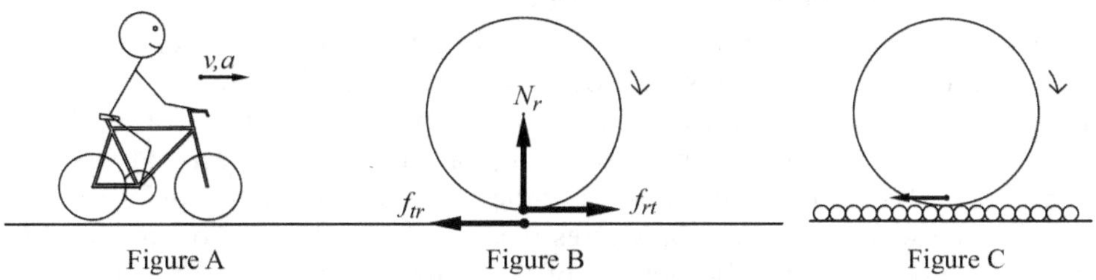

Figure 5.6. (A,B,C) Bike and rider.

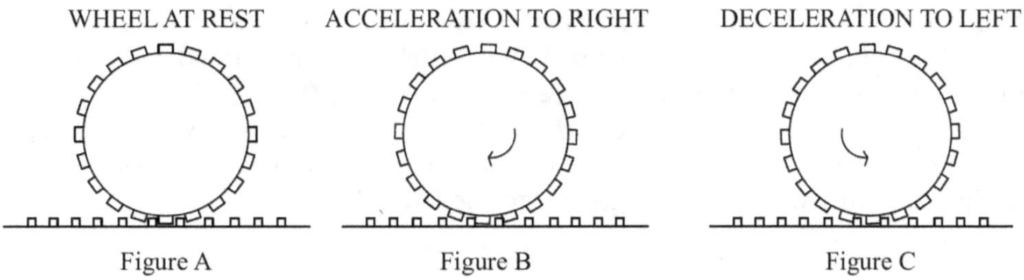

Figure 5.7. (A,B,C) Toothed wheel on ground.

The act of pedaling in the normal fashion causes the rear tire to try to spin in a clockwise sense. In an action very similar to walking, the rear tire pushes on the road toward the left with a force of friction f_{tr} (figure 5.6B). According to Newton's Third Law, the road pushes back on the tire to the right f_{rt}. The ground also pushes vertically upward with the *normal* force.

It might be helpful to visualize the action of f_{tr} by imagining that the ground is covered with small marbles; as the tire tries to spin, the marbles are ejected toward the left (figure 5.6C). The action–reaction pair of forces between the tire and the ground is difficult to visualize. It is especially hard to see the frictional interaction; consider the tire–ground interaction in a series of simple diagrams.

First, examine starting the machine from rest. In order to analyze the role of friction in accelerating the bicycle, consider what occurs as we begin to propel the machine to the right; remember the interlocking rough surfaces as the cause of friction. The rough surfaces are represented as a geared wheel meshed with a toothed ground (figure 5.7A).

As the wheel is given a clockwise spin, the bottom of the tire is trying to move to the left (figure 5.7B). Friction between the tire and the road resists the impending motion of the wheel—note the collision of teeth. Since the tire pushes on the ground toward the left, the ground exerts a rightward force on the bottom of the wheel—preventing it from rotating. Thus, the overall machine is accelerated to the right.

It is this force from the ground, caused by static friction, that produces the forward acceleration of the bicycle!

The statement made earlier that friction opposes motion is still true. In this case, the motion opposed is the rearward slippage at the bottom of the tire.

5.9.2 Acceleration from an existing velocity

Once we understand the actions that result in the initial acceleration of the machine, we need to consider the events that occur once the bicycle is traveling. A wheel, rolling along the ground toward the right, has two distinctive types of motion:

The wheel is *translating* from left to right; call this the translational velocity v_t. The wheel is also *rotating* about its axle; the rotation will be clockwise as seen in figure 5.6. This rotational motion causes the top of the wheel to be moving to the right with velocity v_r, and the bottom of the wheel to be moving to the left with velocity v_r. These motions are *separate actions*—if the wheel was skidding on ice, it could have translational motion but no rotation. If the wheel was suspended in the air (or spinning on the ice), it could have rotational motion but no translation. When a wheel is rolling along the ground, it has both the translational and rotational motions that are equal and opposite vectors.

The motion of the wheel's bottom is of particular interest. When rolling, the bottom has a velocity v_{total} that is the vector sum of the translational component to the right and the rotational component to the left.

$$v_{total} = v_t - v_r = 0$$

The vector sum at the bottom of the wheel is zero; the bottom of the wheel is not moving! There is a cancelation between the translational component and the rotational component. This statement might be surprising—although the bike is traveling to the right, the bottom of the wheel is at rest with respect to the ground.

Suppose a force is applied to the pedal; the wheel tries to rotate clockwise a bit faster. The increased rotational contribution to the velocity at the bottom of the tire results in a vector sum that is no longer zero. As the tire is trying to move backward, it pushes leftward on the ground with the force of static friction. From Newton's Third Law, the ground pushes forward on the bicycle (figure 5.7B). This forward force from the ground is the net external force that accelerates the already moving bicycle.

5.9.3 Deceleration from an existing velocity

How does the act of braking—internal forces between the bicycle's components—lead to a stopping of a bicycle?

How about braking? Regardless of the type of brakes (rim caliper, disk, coaster, etc), the action of braking decreases the rotational motion of the wheel. The bottom of the wheel now has a net velocity that is no longer zero; the wheel pushes on the ground to the right. Consequently, Newton's Third Law states that the ground pushes back on the wheel to the left. This leftward push from the ground is the *external force of deceleration.*

It might be helpful to envision the toothed wheel. As the wheel attempts to spin more slowly, the wheel's teeth collide on their right side. The teeth from the ground exert a force on the wheel toward the left (figure 5.7C).

5.10 Maximum acceleration of a bicycle

5.10.1 Maximum acceleration

One of our youthful instincts when we hop on a bike is a desire to go fast; let us think about the maximum acceleration possible on a bicycle.

The force of static friction produces the bike's acceleration; hence, the maximum force and, therefore, the maximum acceleration, is determined by the magnitude of the static frictional force. To achieve maximum acceleration the rider must not spin the rear wheel. When the wheel is spinning, sliding friction applies; the coefficient and force of sliding friction is less than the coefficient and force of static friction. The maximum force of acceleration occurs when the rider pedals just hard enough to put the rear wheel on the verge of spinning. The force of static friction is expressed:

$$f_{max} = \mu_{static} \times N_r$$

where:

- μ_{static} is the coefficient of static friction, typically $\mu_{static} = 0.9$ for rubber on a concrete road.
- N_r is the vertical, upward force on the rear tire (normal force). In the typical seated riding position, N_r is probably a bit more than one-half (about 60%) the weight of the bike and rider.

We can calculate the maximum acceleration of the bicycle caused by the largest possible force of friction. Our standard rider of 161 lb on a 20 lb bike results in a system mass of 5.62 slugs that must be accelerated. The normal force at the rear wheel carries sixty percent of the system's weight:

$$N_r = 0.6 \times 181 \text{ lbs} = 109 \text{ lb}$$

The maximum value of the frictional force is:

$$f_{max} = \mu\, N_r = 0.9 \times 109 \text{ lb} = 98.1 \text{ lb}$$

Thus, his maximum acceleration is:

$$a = \frac{f}{m} = \frac{98.1 \text{ lb}}{5.62 \text{ slugs}} = 17.5 \text{ ft/s}^2 = 0.54 \text{ g}$$

This result should not be surprising—a force, equal to 54% of the system's weight (force of the earth's gravity), produces an acceleration equal to 54% of gravity's acceleration!

The above calculation is based on a seated rider—greater forces could be generated if the rider shifted more weight toward the rear of the bike. While he cannot change the system weight of 181 lb, he could create a situation in which 80% of the system weight is on the rear tire and therefore increase the maximum acceleration.

How realistic is the above situation? Realize that it is a calculation based on the maximum force of friction with a coefficient near unity. There may be additional considerations.

Can any rider pedal so strongly that the maximum force of friction is the primary consideration in the acceleration of the bicycle?

The relationship, borrowed from chapter 11, for a mountain bike in low gear shows that the force on the pedal is related to the force on the road by:

$$F_{pedal} = 1.22 \times F_{road}$$

a road force of 109 lb requires a pedal force of:

$$F_{pedal} = 1.22 \times F_{road} = 1.22 \times 98.1 \text{ lb} = 120 \text{ lb}$$

If the rider were to rise from the saddle and put most of his weight on one pedal, a force of 120 lb is not difficult to attain.

What are the consequences of a rider applying a large force to the pedal?

While it is easy to generate a large force of acceleration, we must be careful not to always assume that the outcome is, indeed, a large forward acceleration. The rider/bike system is not a simple point mass (concentrated in one spot); it is an extended body capable of rotational motions. *There might be trouble with these rotations!*

5.10.2 Maximum deceleration

As with acceleration, deceleration is determined by the friction grip between the tires and the road.

$$a = \frac{f}{m}$$

with the maximum force of friction $f = \mu \times N$:

$$a_{max} = \frac{\mu \times N}{m}$$

When the bike is stopping, both wheels can be used; the normal force is the full system weight of 181 lb.

- for static friction between rubber and concrete $\mu_{\text{static}} = 0.9$

$$a_{\text{max no skid}} = \frac{0.9 \times 181 \text{ lb}}{5.62 \text{ slug}} = 29 \frac{\text{ft}}{\text{s}^2}$$

- when the bike is skidding, the much lower coefficient of sliding friction applies; let us use $\mu_{\text{sliding}} = 0.45$

$$a_{\text{max with skid}} = \frac{0.45 \times 181 \text{ lb}}{5.62 \text{ slug}} = 14.5 \frac{\text{ft}}{\text{s}^2}$$

Since a skidding deceleration is one half, the stopping distance will be doubled!

An additional complication of hard stops is the phenomena of the header (chapter 11).

5.11 Velocity and acceleration of a bicycle

For several reasons, the maximum acceleration of the previous section may not be feasible. These reasons include factors such as the strength of the rider, the danger of wheelies, reduced friction, etc. In this section, we consider accelerations that are more realistic. In this initial analysis, we will ignore resistive forces such as air resistance and rolling resistance. Later, we shall see that these resistive forces are very small at low speeds. We continue to use a 161 lb rider on a 20 lb bike and a mass of 5.62 slugs. Perhaps our rider is not exerting her maximum effort. In a leisurely 'ride in the park', the typical cyclist's effort might produce an acceleration force that is only ten percent of the system weight.

There may be more beautiful times, but this one is ours.

Jean-Paul Sartre

The force of acceleration will be: $F = 0.1 \times w = 0.1 \times 181 \text{ lb} = 18.1 \text{ lb}$

For the acceleration, we get:

$$a = \frac{F}{m} = \frac{18.1 \text{ lb}}{5.62 \text{ slugs}} = 3.22 \text{ ft/s}^2$$

How long does it take to attain a speed of 5 mph (7.33 ft/s)? From chapter 4, we have the expression for velocity with constant acceleration: $v_f = at + v_i$

If the rider starts from rest: $v_i = 0$, our equation becomes: $v_f = at + 0 = at$ rearranging and substituting numbers in:

$$t = \frac{v_f}{a} = \frac{7.33 \text{ ft/s}}{3.22 \text{ ft/s}^2} = 2.28 \text{ s}$$

How far has our rider traveled before she hits 5 mph? From chapter 4, our expression of distance traveled with constant acceleration:

$$d = v_i t + \frac{1}{2}at^2 = \frac{1}{2} at^2 = \frac{1}{2}(3.22 \text{ ft/s}^2)(2.28 \text{ s})^2 = 8.37 \text{ ft}$$

What happens if our cyclist maintains the pedaling effort? The accelerating force of 18.1 lb resulted in an acceleration of 3.22 ft/s^2 and it only took her 2.28 s to hit 5 mph (7.33 ft/s). Will she reach 10 mph in the next 2.28 s? How about 20 mph in 9.12 s? How about 50 mph in 22.8 s? Why not 500 mph in 228 s? For better or worse, this will not happen. The reason is that from 0 to 5 mph the resistive forces are small. The primary horizontal force on the rider is the forward force causing the acceleration. There are, however, resistive forces opposing the motion of the bicycle. The next section examines in detail the resistive forces.

5.12 Resistive forces on a moving bicycle

Why, in light of Newton's First Law of Motion, do we have to keep pedaling to travel at a constant velocity?

Newton's First Law of Motion states that, in the absence of external forces, an object in motion at a constant velocity will remain in motion at that constant velocity. The traveling cyclist encounters a variety of external forces opposing the motion of the system. If the rider wishes to travel at a constant velocity, she must apply force to the pedals. The force on the pedals ultimately results in the ground applying a forward force to the bicycle. When the forward force is equal to the total rearward forces of resistance, the machine travels at a constant velocity.

What are the forces opposing the motion of the bike?

The resistive forces on a traveling bicycle are caused by various physical phenomena. The primary forces of resistance are:

- gravitational resistance on hills
- air resistance

- rolling resistance of the tires
- bearing resistance in hubs of wheels

The first force in our list, the *resistance of gravity*, is due to the weight of the system—specifically, the component of gravity that is parallel to the hill. Gravitation effects are explored in chapter 6. The last three forces are listed in the order of their relative magnitude and consequent impact on the bike's motion. There are also other internal but small resistive forces due to chain rollers, etc [2].

5.13 Air resistance

How significant is the effect of air resistance?

You ride by so fast on the headlong blast.

Faust,ii Shelley

To get a feeling for the action and importance of air resistance, we need to think more about air. Air might appear to have an insignificant mass and weight; after all, we push it aside all day as we walk about. There are several points to ponder.

Our typical walking encounter with air occurs at slow speed—even a fast walk or jog might be at a speed of only a few mph. The impact of air resistance dramatically increases with the speed of the moving object. You can readily observe this effect by walking at 4 mph holding out your hand; the force of air resistance is imperceptible. In a car moving at 40 mph, hold out your hand through an open window. At 40 mph, the force of air resistance on your hand will be 100 times as large! For an even more dramatic effect, cut a piece of foam board to the same area as the frontal area of a rider, about 30 inches on a side. While someone else is driving (on an isolated road with no traffic), hold the foam board out the car window with the board's broad surface facing the wind. As the car travels at various speeds, the substantial forces you experience are equivalent to the force on a cyclist traveling at the same speeds.

The reason for large air resistance forces is that air is actually *quite heavy and massive*. The density of air at 60 °F and sea level is $\rho = 0.00238$ slugs/ft^3. Recall from section 3.8 that a beach ball, four foot in diameter, has a weight of 2.56 pounds. Air has a significant mass. In order to derive a simple expression for the force of air resistance, think of a cube of air (figure 5.8):

As a cyclist moves through air, consider the air as an obstacle that must be pushed aside. If m is the mass of the air inside the cube, the required force to accelerate this air is given by Newton's Second Law:

$$F = ma \tag{5.3}$$

The air is initially at rest; suppose it is accelerated up to the speed of the bike and rider.

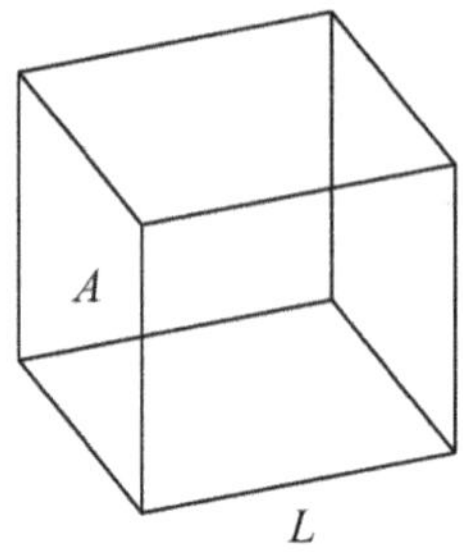

Figure 5.8. Cube of air.

$$a = \frac{\Delta v}{t} = \frac{v - 0}{t} = \frac{v}{t} \tag{5.4}$$

With ρ for the density of air and V for its volume (volume equals area A times length L)

$$\rho = \frac{m}{V} = \frac{m}{A\,L} \tag{5.5}$$

rearranging:

$$m = \rho AL \tag{5.6}$$

substituting (5.4) and (5.6) into (5.3), we obtain:

$$F = (\rho AL)\left(\frac{v}{t}\right) = \rho Av\frac{L}{t} \tag{5.7}$$

In equation (5.7) L/t is the distance we must push the air divided by the time. This is simply v, the speed of the rider. Thus:

$$F_{\text{air}} = \rho Avv = \rho Av^2 \tag{5.8}$$

Spend a few minutes and consider how each of these factors impacts air resistance. As expected, the density of the air is important; thinner air at high altitudes is easier to push aside. Also, the cross sectional (frontal) area of the bike and rider A is very significant. Do you see why many riders like to crouch down and, therefore, reduce the cross sectional area they present to the air? Although we derived this expression for a cube shaped object, the actual shape of the cross sectional area is not significant. The speed of the rider enters as a squared term. The reason is that the velocity of the rider determines the depth L of the cube; it effectively determines how far the rider must move the air to get it out of the way. In addition, the speed of the rider determines the acceleration that must be imparted to the air.

In some ways, the above derivation was overly pessimistic. We postulated that the air had to be 'picked up' and accelerated to the cyclist's speed. As an analogy, envision a football player clearing a path for the ball carrier. Our derivation was equivalent to asking the blocker to grab the opposing player and bring him up to the blocker's speed. No doubt, picking up opposing players and carrying them along will invoke a penalty, but it is also a huge waste of effort. Since the blocker merely needs to clear the path for the runner, it is only necessary to push the tacklers aside. In a similar manner, the bike rider needs only to push the air aside. A more detailed analysis of air resistance results in the same basic equation we have derived—only reduced by a factor of fifty percent. Thus:

$$F_{air} = 0.5\,\rho A v^2 \tag{5.9}$$

A complete analysis of air resistance must examine streamlining effects that result in a further reduction beyond this fifty percent. Also, objects with long thin shapes encounter skin friction along their length. Further complications such as turbulence occur when the air speeds are high. We will, however, make use of the simple expression in equation (5.9) and find good agreement with real world data.

In order to do calculations of air resistance in U.S. Customary units, the density of air must be in slugs/cubic feet and the cross sectional area of the rider in square feet. The speed of the rider must be in feet/second and then the calculated force will be expressed in pounds. A sample calculation is straightforward with the following typical values:

- the density of air: $\rho = 0.00238\frac{\text{slugs}}{\text{ft}^3}$
- the area of our standard rider: $A = 5.38\ \text{ft}^2$
- the rider's speed: $v = 20\ \text{mph} = 29.3\frac{\text{ft}}{\text{s}}$

Results in a force of air resistance:

$$F_{air} = 0.5\,\rho A v^2 = 0.5 \times \left(0.00238\frac{\text{slugs}}{\text{ft}^3}\right) \times (5.38\ \text{ft}^2) \times \left(29.3\frac{\text{ft}}{\text{s}}\right)^2 = 5.50\ \text{lb}$$

Sometimes it is useful to write:

$$F_{air} = 0.5\,\rho A v^2 = k v^2 \tag{5.10}$$

where we have grouped the one-half, the air density and frontal area terms into a constant term k:

$$k = 0.5\,\rho A = 0.5 \times \left(0.00238\frac{\text{slugs}}{\text{ft}^3}\right) \times (5.38\ \text{ft}^2) = 0.00640\frac{\text{slugs}}{\text{ft}} \tag{5.11}$$

hence

$$F_{air} = 0.00640\,v^2 \quad \text{(speed in ft/s)} \tag{5.12}$$

Although it simplifies calculations, one must be careful in using this value of k since it is critically dependent upon: the size and riding position of the rider (frontal area, A), the density of the air (ρ), the extent of aerodynamic streamlining (0.5), and the units selected for the calculation.

Since most riders in the United States think of their speeds in miles per hour, it is useful to work with an adjusted k parameter. One mph is 1.47 ft/s; we can adjust the k factor by 1.47^2 to obtain:

$$F_{air} = 0.0138\, v^2 \quad \text{(speed in mph)} \tag{5.13}$$

A few sample calculations demonstrate the impressive impact of air resistance on a moving cyclist.

at 5 mph, $F_{air} = 0.0138\, v^2 = 0.0138 \times (5\text{ mph})^2 = 0.345$ lb
at 10 mph, $F_{air} = 0.0138\, v^2 = 0.0138 \times (10\text{ mph})^2 = 1.38$ lb
at 20 mph, $F_{air} = 0.0138\, v^2 = 0.0138 \times (20\text{ mph})^2 = 5.52$ lb
at 40 mph: $F_{air} = 0.0138\, v^2 = 0.0138 \times (40\text{ mph})^2 = 22.1$ lb

Note: there will be insignificant round off discrepancies when comparing calculation done in ft/s and mph.

The force of air resistance at 20 mph is *four times* that at 10 mph. The force goes as the square of the speed. If you double the speed, the force is quadrupled. These forces of air resistance are dramatic and highly significant to the cyclist. Can speeds over 40 mph be attained on a bicycle? Yes, professional racers going down hills hit such speeds. What about the rest of us Sunday afternoon riders? We might not hit 40 mph but we do encounter winds. Have you ever noticed that it can be very difficult to ride into a head wind? The reason is that the force of air resistance depends upon the relative speed between the rider and the air. If you are riding at 15 mph (ground speed) and encounter a head wind of 10 mph, the force of air resistance is equivalent to riding at 25 mph! Riding at 20 mph into a strong head wind gust of 20 mph results in an air resistance equivalent of riding at 40 mph.

Her hardy face repels the tanning wind.

Health, Thomas Parnell

The above discussion has concentrated on the force that must be exerted to overcome the air resistive force on the moving bicycle. Chapter 8 examines concepts of energy and power. Although the concept of force is certainly important in understanding the motion of the bicycle, the concept of power, specifically the power output of the rider, is the primary consideration in understanding the overall effort required to propel a bicycle over a measurable period of time.

A more sophisticated and mathematically advanced analysis is needed to understand issues such as: laminar/non-laminar flow, Reynolds numbers, and drag coefficients. In world-class competitive racing, the racers wear clothes that have near microscopic surface textures. The fabric is designed to control the airflow across their bodies. Additional consideration should also be given to the effect of air resistance on the spinning spokes!

In all of the calculations, it is possible to lose grasp of the very significant impact of the cyclist's effort in fighting air resistance while traveling at high speeds. Attempts have been made to cancel the effect of air resistance by drafting fast moving objects. The results have been phenomenal speeds. One of the earliest was on June 30, 1899 when Charles M Murphy drafted a train on the Long Island Railroad. He rode on boards placed across the ties. To shield himself from the external mass of air, Murphy was shrouded within a hood attached to the last car. On a racing bicycle, Murphy was able to keep up with the train as it traveled one mile in 57.8 seconds—better than 60 mph! He earned the nickname 'Mile a Minute Murphy'. Another astonishing speed record was set in 1986 by John Howard. By drafting a racecar on the ultra-smooth Bonneville Salt Flats, Howard was able to attain a speed of 152.28 mph.

5.14 Rolling resistance

Earlier in this chapter, we discussed pushing a box across a floor. Sliding the box involved resistive forces of static and sliding friction. Suppose we make the job easier by putting the box on wheels and simply roll the box along the floor. No doubt, it is easier to push the box on wheels; however, the rolling wheels still offer some resistance—the so-called *rolling resistance*. Intuition will lead us to suspect that a hard tire on a hard surface such as a railroad wheel on a steel track will have minimal rolling resistance. Whereas, a soft tire on a spongy surface encounters a high rolling resistance.

If the rolling resistance of a tire is usually associated with the tire's softness or flatness—how can the tendency of a tire to flatten in response to a vertical force lead to a force that impedes horizontal motion?

We have seen in an earlier section of this chapter that horizontal decelerations are caused by horizontal forces. Vertical forces produce changes in vertical velocities. It is not obvious at first how the vertical forces, associated with a soft tire and/or a soft surface, impede horizontal motion. We begin by looking at a few simple diagrams to gain insight into the physics of rolling resistance.

Figure 5.9A illustrates a hard tire at rest on a soft surface. All of the deformation is happening to the ground. As the wheel begins moving to the right, it will press the

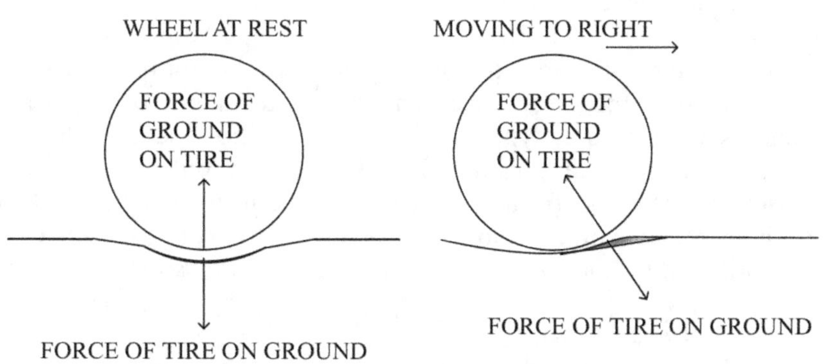

Figure A Figure B

Figure 5.9. (A,B) Hard wheel on soft surface.

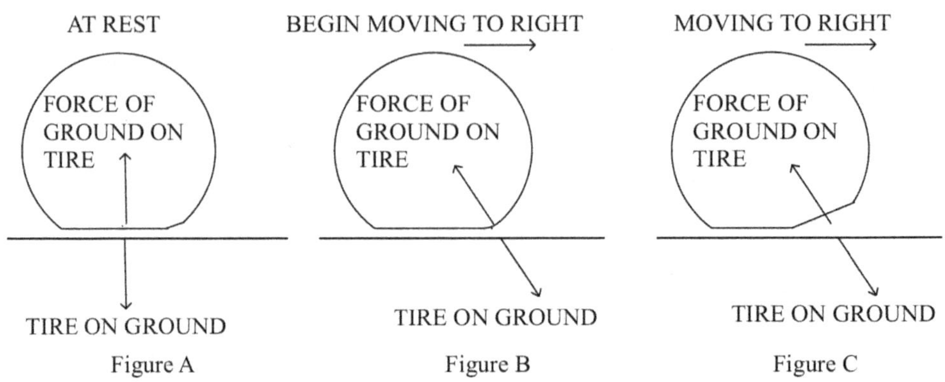

Figure 5.10. (A,B,C) Soft wheel on hard surface.

ground on the right side (figure 5.9B). By Newton's Third Law, the ground responds with an equal and opposite force. There is a horizontal component to these vectors. *The horizontal component of the ground on the tire is the force of rolling resistance.*

Many cases of rolling resistance, such as a pneumatic bike tire on a road, involve a soft tire on a hard surface (figure 5.10A).

Once the wheel starts moving, the ground pushes on the tire and flattens the bottom right edge of the tire (figure 5.10B and C). Again, there is a horizontal component to the force from the ground. If we were dealing with ideal elastic materials that had no energy losses, then the bounce back at the left trailing edge would give a push to the right and return the energy to the wheel. Unfortunately, real materials and surfaces do not possess such perfect elasticity. The introduction of pneumatic tires in the 1890s afforded a substantial reduction in rolling resistance when compared to solid rubber tires. The air within the pneumatic tire acts as a spring with reduced losses between the compression on the tire's leading edge and the expansion that occurs on the trailing side. In comparison to a solid rubber tire, a pneumatic tire has less rubber to flex. This flexing of the tire's fibers results in a higher rolling resistance.

Later in this book, we will discuss a method that can be used by the reader to determine the actual resistive forces on a particular bicycle. For now, we make use of published values of rolling resistance. The rolling resistance of a particular tire depends upon such factors as type of rubber, construction method, thread design, and, of course, inflation air pressure. Professor David Wilson quotes values of rolling resistance that range from approximately one to four newtons of force per tire [3]. An intermediate value is about 4.5 newtons for two tires. In U.S. Customary units, this works out to approximately *one pound of rolling resistance* per bicycle.

$$F_{\text{rolling}} = 1.0 \text{ lb} \tag{5.14}$$

At this juncture, it is worth appreciating how small this rolling resistance really is. Suppose, instead of rolling, the wheels were slid along—in our discussions on static and sliding friction (sections 5.6 and 5.7), we saw the coefficients of friction for rubber on pavement range from 0.5 to 4.0. Using the lowest value of 0.5 for sliding friction will yield a resistive force on a 181 lb rider/bike system of:

$$f = \mu N = 0.5 \times 181 \text{ lb} = 90.5 \text{ lb}$$

Thus, the act of rolling a body on wheels requires an effort of approximately *one percent of the force* when compared to sliding the same weight. Think back to those cave dwellers we discussed in chapter 1; it is a lot easier to roll that rock than it is to slide it. *The magic of the wheel!*

5.15 Bearing resistance

What is the advantage of using ball bearings to support moving parts?

Strictly speaking, the resistance within the wheel hubs is not external to the bike/rider system. The bearing resistance plays a similar role as the brake pads acting on the rims. As the rotational speed of the wheel is reduced, the frictional force from the ground serves as the external force. Nevertheless, the bearing losses can be modeled as a resistive force against the bike's motion.

Ball bearings are commonly used at various locations in the bicycle for the simple reason that the resistive forces are due to rolling resistance rather than sliding resistance. Recall our caveman rolling the large rock rather than trying to slide it. For a more contemporary image, think about sliding a heavy box across a floor. The force of static friction must be overcome to get the box in initial motion and the force of sliding friction to continue the motion. The chore would be much easier if you had placed thousands of ball bearings under the box and rolled it over the bearings.

In the previous section, we saw the wheel's advantage to be a direct consequence of rolling resistance being so much smaller than sliding friction. Rolling resistance is attributed to the deformation of the contact surfaces. The harder the contact surfaces, the less the deformation; thus, the rolling resistance is smaller. Steel ball bearings contained within a steel casing will have minimal deformations with very small rolling resistance. In a wheel axle, supported between the bearings of the hub, the ball bearings are able to roll as the axle turns. This arrangement offers far less resistance than if the axle were turning in a tight bushing. The resistive forces of the bushing would be that of sliding friction. Cylindrical bearings function in much the same way as ball bearings.

Published values for the various resistances of steel on steel are as follows:

steel sliding on steel

$$\mu_{\text{sliding}} = 0.6$$

steel rolling on steel

$$\mu_{\text{rolling}} = 0.002$$

Thus, sliding friction is approximately 300 times greater than rolling resistance [4]. In a bicycle, bearings are found in many locations such as the wheel hubs, bottom bracket, fork, pedals, etc. Another very important advantage in the use of bearings is that they are designed as the parts that wear the quickest, but are inexpensive and simple to replace.

For a typical value for the bearings resistance of two bicycle wheels, we use: [5]

$$F_{\text{bearing}} = 0.006 \text{ lb} \tag{5.15}$$

Note: this value is insignificant in comparison to the other resistances. Since the other resistances are given to a lower precision, the mathematically exact have reason to complain. We are including it for the sake of completeness.

The resistive forces on the bicycle may be written as the sum of constant values of rolling resistance and bearing resistance and an air resistance term that is proportional to the square of the speed:

$$F_{\text{tot}} = F_{\text{rolling}} + F_{\text{bearing}} + F_{\text{air}} = F_{\text{rolling}} + F_{\text{bearing}} + kv^2 \tag{5.16}$$

A slightly more compact expression results if we combine the constant terms of bearing and rolling resistance into a single variable F_{rb} and refer to the overall sum as F_{tot}:

$$F_{\text{tot}} = F_{\text{rb}} + kv^2 \tag{5.17}$$

Table 5.2. Forces of air + rolling + bearing resistances.

Air resistance constant = rolling 0.0138, speed in mph, area = 5.38sq ft				
Speed (mph)	Air resistance (lb)	Rolling (lb)	Bearing (lb)	Total (lb)
2	0.06	1.0	0.006	1.06
4	0.22	1.0	0.006	1.23
6	0.50	1.0	0.006	1.50
8	0.88	1.0	0.006	1.89
10	1.38	1.0	0.006	2.39
12	1.99	1.0	0.006	2.99
14	2.70	1.0	0.006	3.71
16	3.53	1.0	0.006	4.54
18	4.47	1.0	0.006	5.48
20	5.52	1.0	0.006	6.53
22	6.68	1.0	0.006	7.69
24	7.95	1.0	0.006	8.95
26	9.33	1.0	0.006	10.3
28	10.8	1.0	0.006	11.8
30	12.4	1.0	0.006	13.4
32	14.1	1.0	0.006	15.1
34	16.0	1.0	0.006	17.0
36	17.9	1.0	0.006	18.9
38	19.9	1.0	0.006	20.9
40	22.1	1.0	0.006	23.1

Note: the mathematical step of rounding the total force to one or two decimal places has the effect of ignoring the bearing resistance.

We can add the individual contributions of rolling (equation (5.14)), bearing (equation (5.15)), and air resistance (equation (5.13)) to construct a table of total resistive force at various speeds. The values of rolling and bearing resistance are constant with respect to the speed of the bicycle (table 5.2).

> The dominance of air resistance is compelling; at all but the lowest speeds (8 mph and below), air resistance is the most significant force opposing the motion of the bicycle. Perhaps, it is best to consider the situation from a positive perspective. The advances in designs and materials have resulted in two-wheelers that operate with minimal forces of bearing and tire resistances. It is unlikely that scientific and engineering advances will eliminate the need for air on the planet. Of course, great efforts are made in rider position, clothing, fairings, etc, to minimize the consequences of having to plow through the atmosphere.

5.16 Coasting—simplified analysis

One of the really fun aspects of cycling is a consequence of the bike's ability to freely coast for long distances. Just as Galileo studied motion in a simplified manner, we will do the same. Dealing with forces of bearing and rolling resistance is straightforward. These forces are constant with normal riding speeds; Newton's three laws provide an easy analysis. Since air resistance is a function of the speed, motion analysis including air resistance requires a more complex, calculus based approach. In this section, we evaluate the motion of the cyclist in the absence of air resistance (Galileo did the same). A rolling resistance of 1.0 lb and a bearing resistance of 0.00628 lb result in a force of 1.00628 lb. If we proceed further with the use of these numbers, we face a serious breach of rules regarding precision in mathematical calculations. It is mathematically wrong to simply add these two numbers and pretend we know the total resistance to the five decimal places. The sum should be rounded to just the rolling resistance. The problem is that it would be easy to lose this '1.0' (and totally forget about bearing resistance) in the calculations. Therefore, we will show the total resistance as 1.006 lb.

If this were the only force slowing the bike, how far could we coast from a top speed of 20 mph (29.3 ft/s)? Using Newton's Second Law, we get the value for the deceleration of the bike/rider system of mass 5.62 slugs:

$$a = \frac{F}{m} = \frac{-1.006\text{ lb}}{5.62} = -0.179\text{ ft/s}^2$$

The minus sign reflects the fact that the force is in opposition to the direction of motion; hence, it is slowing down the bike. This analysis is similar to the acceleration calculation performed earlier in this chapter; only now, there is a force of deceleration rather than acceleration.

For the final velocity with constant acceleration: $v_f = at + v_i$
If the rider ends at rest: $v_f = 0$
Our equation becomes: $0 = at + v_i$
Upon rearranging and substituting values the time of the coast is:

$$t = \frac{-v_i}{a} = \frac{-29.3\ \frac{\text{ft}}{\text{s}}}{-0.179\ \frac{\text{ft}}{\text{s}^2}} = 164\ \text{s}$$

A coast of almost three minutes! How far has our rider traveled before he comes to rest from the 20 mph?

From chapter 4:

$$x = v_i\,t + \frac{1}{2}at^2$$

Substituting numbers: $x = 29.3 \times 164 + \frac{1}{2}(-0.179)(164)^2 = 2400\ \text{ft}$. A nice long coast!

This discussion did not include the force of air resistance since it varies with the speed of the cyclist. A non-constant deceleration requires the calculus which is beyond the scope of this book. However, we will summarize the results of a coast from 20 mph subject to the forces of air resistance, rolling resistance, and bearing resistance as follows: time of coast = 82 s and distance of coast = 820 ft—still nice!

5.17 Force analysis walking versus riding

Why is riding a bicycle more efficient than walking?

Much of the joy of cycling comes from the two-wheeler's ease of movement. On a level, hard pavement the machine glides almost effortlessly along. An easy bike ride might move the rider a distance of ten miles in an hour. Walking, with approximately the same effort, might only cover a distance of a few miles. Yet, both modes of motion—riding and walking—are primarily driven by the body's lower muscles.

In this section, we consider the forces involved in the act of bipedal walking/running and compare the effort to cycling. A force analysis is not always the most suitable approach in analyzing complex mechanical processes such as human walking and running. The human body consists of nearly a thousand moving muscles, tendons, and bones. Forces continuously vary in both magnitude and direction. Many muscle groups act in opposition to one another. In addition, direct measurements of the tensile forces exerted by the muscles are invasive and certainly unpleasant for the experimental subject. Nonetheless, basic physical principles provide insight as to how the bicycle affords such gains in human motive efficiency.

Although bipedal walking preceded cycling in both our evolution and childhood development, an understanding of the mechanics of walking is more complex and difficult to comprehend than propulsion on a bicycle. The exact nature of the bipedal walking/running gait is the subject of extensive modern research [6–9]. While the mechanics of walking might be complex, one fact is certain; there are basic physical principles, Newton's Three Laws of Motion, to guide us. It might appear puzzling why, once the human body is in motion, it is necessary to exert repetitive forces to maintain its motion. The explanation lies in the complex nature of the gait. After the initial step, it is necessary for the body to exert continually a steady series of acceleration, and surprisingly, deceleration forces. The required forces of exertion are much greater than what would be required to overcome common resistive forces such as air resistance.

The First Law implies that no net force is required to travel at a constant velocity. It might appear that the requirement for the human body to constantly exert forces during uniform walking is a violation of Newton's law. The reader may rest assured that the act of walking does not contradict Sir Isaac's or any other laws of physics. From the very active research on the forces and mechanics of walking, we can understand a few key aspects of the human gait.

As the upper body moves, the opposite leg must be swung into motion; another force of acceleration is required to move this leg.

As the body tries to move forward with the first foot pushing backward on the ground, Newton's Third Law tells us that the ground pushes forward on the foot. The foot and connected leg and upper body are accelerated forward.

As the legs swing forward in front of the body, the feet eventually strike the ground. Since the ground is at rest, the forward moving foot must be slowed down (decelerated) to a zero velocity. The body's lower muscles provide this force of deceleration. If the forward swinging limbs were stopped only by contact with the ground, the result would be a significant jar. Thus, a step requires both an acceleration phase as the leg is brought up to speed and a deceleration phase as the leg is brought to rest. Walking a mile requires approximately 2000 steps with a stride of about 30 inches, each step involving the acceleration and deceleration of a leg. This starting and stopping of the large leg masses requires significant and repetitive exertion of forces by the body's lower muscles.

The swinging leg is sometimes compared to a pendulum, and it is commonly observed that pendulums are capable of swinging for long times after an initial push. Why is it then necessary to exert effort to maintain the swinging of the legs? It is a matter of the frequency of the swing. The period, the time it takes a pendulum to swing back and forth, is determined by the length of the pendulum. A grandfather clock running too slow or fast is adjusted by moving the pendulum bob up or down. The human body's pendulum leg has a natural period of about 1.56 s per step or 0.64 swings per second. This slow rate of natural swing along with a stride length of about 30 inches provides a walking speed of only 1.1 mph.

Such a low speed is a stroll. Walking faster, even a few miles an hour, requires us to swing our legs faster than their natural pendulum frequency. If we maintain the same stride length, we would have to swing the leg twice as fast in order to double the walking speed. A brisk walk at four miles per hour requires almost a quadrupling of the legs' natural swing frequency. Thus, the body is forced to swing its legs at what is, in effect, an unnatural frequency. Forcing the legs to swing at their unnatural, resonant frequency requires significant forces from the body's muscles. In addition, once we make the effort and get the leg swinging, we must almost immediately have to exert equivalent forces of deceleration to stop the leg. These muscular efforts, consisting of short bursts of acceleration and deceleration forces—starting and stopping the leg swings, are repeated thousands of times per mile—are not ideal from an efficiency perspective. Cycling with the steady, rotating motion of the lower limbs does not incur these inefficient accelerations/decelerations.

Another aspect of the human gait is that there is a vertical motion to the body's center of mass. As the legs move repeatedly from the vertical to a diagonal position, there is a lowering and raising of the body's center of mass. Although lowering of the center of mass is caused by gravity, it is necessary for the leg muscles to raise the center of mass back up. Cycling on a level road allows the body's center of mass to travel at a constant height. The cyclist does not have to repeatedly raise her center of mass

There is yet another metabolic cost of bipedal propulsion—we must exert vertical forces to support the full weight of the body. Again, the bicycle has the advantage with the saddle supporting a large fraction of the rider's weight.

It is interesting to wonder why the process of evolution has produced a bipedal gait with such fundamental mechanical inefficiencies. We might even speculate why humans did not evolve wheels rather than legs. An obvious retort is that our legs are called upon to do more than propel the body forward. Our legs allow us to move forward, backward, and side-to-side. We jump and hop. We go slow, or fast. We use our leg muscles to help lift heavy weights. Our legs allow us to crawl, sneak up, and climb up and down. Sometimes, our legs act as weapons. Try doing all of these actions with a wheel! In fact, the human bipedal motion is quite good; it allows us to travel long distances—tens of miles on a daily basis. While our gait may not be mechanically perfect, it is well adapted to a range of environments.

Estimates for the force required for walking at 5 mph on a level ground suggests that an average forward force of about eight percent of body weight is necessary [10]. Thus, a 161 lb person needs about 13 lb of forward force. Earlier in this chapter, we saw that, when cycling at 5 mph, the force of air resistance on a rider would only be 0.345 lb; added to a rolling and bearing resistance of 1.006 lb, the net resistive force is only 1.35 lb. While riding at higher speeds drastically increases air resistance, a runner encounters the same atmosphere—in a less compact stance.

Does the mass of the bike not have a metabolic cost? On a level surface, it is only of consequence during accelerations.

5.18 Average versus instantaneous pedal force

All riders know that the pedaling force varies during the crank cycle. When the cranks are near horizontal, the rider is able to press harder than when cranks are vertical. The variation in pedal force, measured in the late nineteenth century, was represented in the 1896 treatise by Archibald Sharp [11] (figure 5.11).

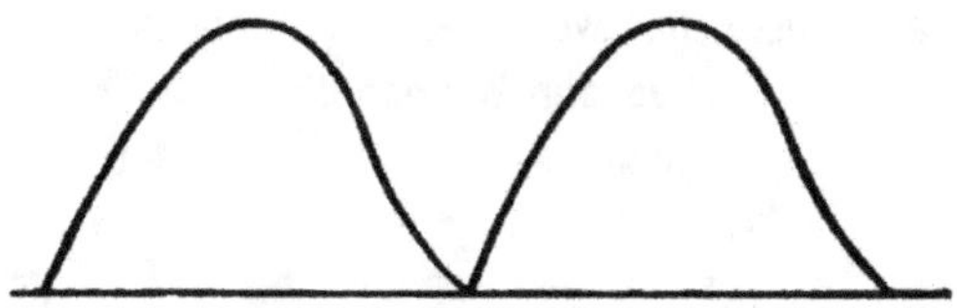

Figure 5.11. Force on pedal Sharp.

The force cycle has a shape familiar in many natural phenomena; mathematically, the curve looks like a common trig function. The average force will be about 65% of the peak force.

References

[1] Lide D R (ed) 1990 *Handbook of Chemistry and Physics* 71st edn (Boca Raton, FL: CRC Press)

[2] Wilson D G 2004 *Bicycling Science* 3rd edn (Cambridge, MA: MIT Press) chapter 6

[3] Wilson D G 2004 *Bicycling Science* 3rd edn (Cambridge, MA: MIT Press) p 211 p 215

[4] Knight R 2008 *Physics for Scientists and Engineers* 2nd edn (San Francisco, CA: Pesarson) p 163

[5] Wilson D G 2004 *Bicycling Science* 3rd edn (Cambridge, MA: MIT Press) p 215

[6] Doke J, Donelan J and Kuo A 2005 Mechanics and energetics of swing of the human leg *J. Exp. Biol.* **208** 439–45

[7] Kuo A 2002 Energetics of actively powered locomotion using the simplest walking model *Biomech. Eng.* **124** 113–20

[8] Taylor C *et al* 1980 Energetic cost of generating muscular force during running—a comparison of large and small animals *J. Exp. Biol.* **86** 9–18

[9] Taylor C 1994 Relating mechanics and energetics during exercise *Comparative Vertebrate Exercise Physiology*vol 38A (New York: Academic) pp 181–215

[10] Wilson D G 2004 *Bicycling Science* 3rd edn (Cambridge, MA: MIT Press) p 159, 215–16

[11] Sharp A 1896 *Bicycles and Tricycles* (London: Longmans) figure 231

Chapter 6

Gravity

Shall gravity cease if you go by?

Essay on Man, Alexander Pope

Gravity is an omnipresent, attractive force that exists between all masses. The most obvious, everyday example is the weight of an object. Weight is the force of gravity between the Earth and another body. If someone 'weighs' 150 lb, it is because the Earth exerts a downward force of gravity of 150 lb on that person. While it is sometimes stated that Isaac Newton 'discovered' gravity, this statement is not quite correct. It might be more accurate to say that the first monkey who fell out of a tree discovered gravity. The earth's force of gravity is pervasive and unavoidable. We become aware of gravity at a very young age. Even a newborn baby is very anxious and apprehensive if not securely held; an aversion to falling is a basic human instinct.

6.1 The basic physics of gravity

If Newton did not 'discover' gravity, what did he do? Why is it called 'Newton's Law of Gravity'? Isaac Newton's insight was to recognize the universal nature of gravity. The legend of the falling apple is a consequence of Newton's perception that it was the same force—*the universal force of gravity*—that pulls the apple to the Earth as the force that reaches into space and pulls on the Moon. The force of gravity on the Moon holds the satellite in orbit about the Earth. Sir Isaac postulated that a force of gravitational attraction exists between any two masses. Newton went further and formulated a mathematical expression that allows for the calculation of the force's magnitude.

Consider any two masses, m_1 and m_2, spaced by a distance R between their centers, figure 6.1A.

Newton stated that the bodies attract each other with a mutual force of gravity, F, and the attraction is proportional to the product of the masses m and inversely

doi:10.1088/978-1-6817-4441-4ch6

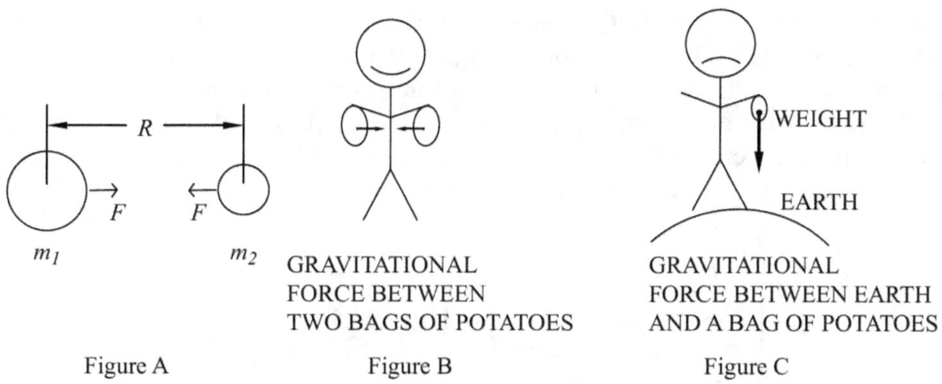

Figure 6.1. (A,B,C) Two masses.

proportional to the square of distance R between their centers. The actual value of the force will depend on the units for mass, distance, and force. Introducing *proportionality constant*, G, called the universal gravitation constant (a fudge factor to make the units come out right), the equation is written:

$$F = G\frac{m_1 m_2}{R^2} \tag{6.1}$$

This equation is called Newton's Law of Universal Gravitation. In U.S. Customary units, where mass is in slugs, distance in feet and force in pounds, the value for G is $3.435 \times 10^{-8}\frac{\text{ft}^3}{\text{slug} - \text{s}}$. An important aspect of Newton's elucidation of gravity is that the same G applies to any two masses. The apple and the Earth, the Moon and the Earth, the Sun and the planets, even objects of cosmic proportion are drawn together by the same law of gravity. The appellation 'Universal Law' is, indeed, appropriate.

If you had two masses, each 1.0 slug, and they are separated by one foot—maybe two 32 lb bags of potatoes one foot apart (figure 6.1B)—the force of gravitation attraction will be:

$$\begin{aligned} F &= G\frac{m_1 m_2}{R^2} \\ &= \left(3.435 \times 10^{-8}\frac{\text{ft}^3}{\text{slug} - \text{s}}\right) \times \frac{(1 \ \text{slug}) \times (1 \ \text{slug})}{(1 \ \text{ft})^2} \\ &= 3.435 \times 10^{-8} \text{ pounds of force} \end{aligned}$$

Only 34 nano-pounds, a very small force! You will not be aware of these sacks pulling on each other. What you would be aware of is how tired your arms feel holding the bags vertically upward.

6.2 Weight of objects

The force of gravitational attraction between two bags of potatoes is negligible. However, if a sack of potatoes is influenced by another very large mass, the gravitational force is quite noticeable. Such is the case when one of the masses is the Earth (figure 6.1C). The force of the Earth's gravitational attraction is called the *weight* of the potatoes. What makes this force of gravity so noticeable is that the Earth has an enormous mass ($m_e = 4.09 \times 10^{23}$ slugs). The distance between the potatoes and the center of the Earth is the radius of the Earth ($R = 2.09 \times 10^7$ ft). The force of gravity on a single bag ($m = 1.0$ slug) is:

$$\begin{aligned} F &= G\frac{m\, m_e}{R^2} \\ &= \left(3.435 \times 10^{-8} \frac{\text{ft}^3}{\text{slug} - \text{s}}\right) \times \frac{(1\ \text{slug}) \times (4.09 \times 10^{23}\ \text{slug})}{(2.09 \times 10^7\ \text{ft})^2} \\ &= 32.2\ \text{pounds of force} \end{aligned}$$

In words, we conclude 1.0 slug 'weighs' 32.2 lb.

With a small rewrite, we easily obtain another useful result; the law of gravity may be written as:

$$F = m\left(\frac{G\, m_e}{R^2}\right) = m\, g$$

and compared to Newton's Second Law $F = ma$

Since F is the gravitational force called weight w and g *is the acceleration of the Earth's gravity*, we obtain:

$$w = mg \tag{6.2}$$

$$\text{where } g = \left(\frac{G\, m_e}{R^2}\right) = 32.2\ \text{ft/s}^2 \tag{6.3}$$

This last relationship is useful in converting back and forth between a body's mass and its weight.

Gravity is a significant factor tiring you out at the end of the day. If you are standing and your weight is 161 lb, it is necessary to push with the feet against the floor with a force of 161 lb. By Newton's Third Law, the ground pushes upward on your feet resisting the tug of gravity.

6.3 'Weight' of object as measured by a scale

Thus far, in this chapter, we have defined weight as the pull of the Earth's gravitational attraction. Confusion develops in that the common use of the word 'weight' is associated with the reading of a scale. Often, it is ok to make this association with the scale. However, what would happen to your scale 'weight' if you stood on the scale inside a brisk elevator? Stand on the scale and note the reading before pressing any elevator buttons; the scale registers your normal weight. Now press up; you will see an increase in the scale reading for as long as the elevator is accelerating upward. There would be a decrease in your scale weight if you had pressed the down button. These increases and decreases in 'weight' only occur during periods of acceleration; normal weigh returns when moving at constant speed between floors.

What if you press the button and the cable breaks? While this is bad news, some very interesting physics can be observed. With a snapped cable, the elevator will be in freefall; you and the scale will also be in freefall. In fact, everything inside the elevator will be falling under the force of gravity. As you look around, the contents of the elevator will appear to be 'floating' about the elevator. If you open a drink and pour out the contents, they will not land on the floor. The liquid will 'float' in front of you. Now while this is all occurring, look down at the scale; it will read zero pounds.

If weight is considered as the scale reading, you are indeed 'weightless'. However, the reason you are crashing into the Earth is that the planet's force of gravity is pulling you down—so you are *not* 'weightless' when weight is defined as the gravitational force.

In this book, the 'weight of an object' means the force of the Earth's gravity.

This broken elevator cable business is no fun. Next time you fly through the air on your mountain bike, you can experience the same effect without the terror of a crashing elevator.

6.4 Force of gravity on a slope—the basic physics

Why is the weight of the bike so important—especially for racing bikes?

If a bicycle is traveling on level ground, the downward tug of gravity is countered by an upward force from the ground and gravity has no effect on the ride. Certainly, in situations of ascending and descending hills, gravity is a major consideration. The pull of gravity is both our foe and friend. Riding downhill, gravity pulls us in the forward direction and acts as a force of acceleration. Traveling uphill, gravity is a resistive force and tries to decelerate the bicycle.

These are certain signs to know faithful friend from flattering foe.

Sonnets to Sundry Notes of Music, VI. 'As it fell upon a day',
William Shakespeare

An object on a sloped surface does not experience the full downward force of gravity. It is necessary to look at the components of the forces as they act parallel and perpendicular to the slope. Figure 6.2 shows a bike rider system on a hill that makes an angle θ with the horizon.

N is the normal force from the sloped ground pressing upward on the bike; it acts perpendicular to the hill. The weight of the system is w; notice this force of gravity pulls vertically downward. The hill's angle θ, is the same as the angle between w and the normal. When riding on significant hills, the biggest force to contend with is the force of gravity—the weight of the bicycle and rider! We will shortly see what it takes to make a hill 'significant'.

Usually when we look at vector components, we consider them in terms of horizontal and vertical directions (x and y). However, in situations involving slopes, the motion occurs along the hill; hence, it is best to resolve vectors into the components that are parallel and perpendicular to the sloped surface. We resolve w into vector components parallel $w_{\parallel}$ and perpendicular $w_{\perp}$ to the surface. The resolution of the vectors is shown on the right side of figure 6.2:

$$w_{\perp} = w \times \cos\theta \quad w_{\parallel} = w \times \sin\theta$$

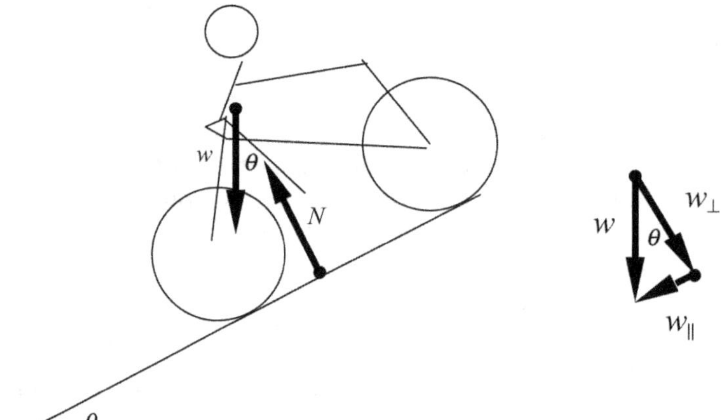

Figure 6.2. Force of gravity on a cyclist on a hill.

These expressions should make intuitive sense: the steeper the hill, the greater the angle, the larger the sine of the angle; therefore, the greater the downward pull of gravity. The component $w_{\|}$, parallel to the hill, is called the *slope force*:

$$F_{\text{slope}} = w_{//} = w \times \sin\theta \qquad (6.4)$$

The slope force always points down the hill. If the bike is initially traveling up the slope, the slope force is a foe—in opposition to the direction of motion. It is referred to as the *slope resistance*. However, if the rider were heading downhill, the slope force is a friend in the direction of motion; it is a force of acceleration.

From chapter 3 equation (3.2), the connection between the slope percentage of a hill and its angle is:

$$\text{slope} = \tan\theta \times 100\% \qquad (6.5)$$

As an example, we obtain the slope force on a 3% hill; first get the angle (using a calculator for max accuracy on inverse tangent):

$$3\% = \tan\theta \times 100\%, \qquad \tan\theta = 0.03, \qquad \theta = 1.718^\circ$$

With rider/bike of weight of 181 lb on this 3% grade, the slope force is then:

$$F_{\text{slope}} = w \times \sin\theta = 181 \text{ lb} \times \sin 1.718^\circ = 5.43 \text{ lb}$$

To ride up the hill at a constant speed, it is not necessary to fight the entire 181 lb force of gravity, rather the much smaller slope force of only 5.43 lb.

Obviously, with steeper grades the slope force is much greater; table 6.1 lists slope forces for our standard rider on a range of hill slopes.

Table 6.1. Slope force on 181 lb rider/bike system for various slopes.

Slope	Angle	Slope force
1%	0.573°	1.81 lb
2%	1.146°	3.62 lb
3%	1.718°	5.43 lb
4%	2.291°	7.24 lb
5%	2.862°	9.04 lb
6%	3.434°	10.8 lb
7%	4.004°	12.6 lb
8%	4.574°	14.4 lb
9%	5.143°	16.2 lb
10%	5.711°	18.0 lb
15%	8.531°	26.9 lb
20%	11.31°	35.5 lb
25%	14.04°	43.9 lb
30%	16.70°	52.0 lb

6.5 Riding uphill at a constant speed

In previous sections, hills were viewed as terrain to coast up and down. On an average ride, very few slopes can be conquered through one long coast. Normally, as we ride along and encounter a hill, our first instinct is to try to maintain speed as we pedal up the hill. In this scenario, the slope resistance represents the additional force that must be developed from increased pedaling effort. For instance, a 10% slope would require an additional forward force on the 181 lb bicycle/rider system equal to the slope resistance of 18.0 lb.

Although maintaining a constant speed while riding up a hill can be physically challenging, it is an easier situation to analyze from a physics perspective. If you struggle and maintain a steady uphill speed, the air resistive force is constant and the math is considerably easier. Therefore, if you are willing to ask your body to pedal hard enough, you can give your brain a break. A rider traveling at a constant speed up a hill must overcome the sum of the slope resistance F_{slope} and the other resistive forces $F_{\text{resistive}}$. In table 5.2, we saw that a bicycle traveling at 20 mph experiences a net resistive force of 6.53 lb due to rolling, bearing, and air resistances. Maintaining a constant speed of 20 mph going up the 10% slope requires a pedaling effort that produces a forward force equal to the sum of the opposing forces:

$$F_{\text{slope}} + F_{\text{resistive}} = 18.0 \text{ lb} + 6.53 \text{ lb} = 24.5 \text{ lb}$$

Is 24.5 lb such a big deal for our strong and fit 161 lb rider? Well, this 24.5 lb is the net external forward force exerted on the bicycle by the ground. We will fully discuss the gearing system in chapter 11 and will learn that the bicycle's gearing system usually operates at a poor mechanical advantage. For now, we can jump ahead to chapter 11 (table 11.1) and borrow the fact that, for a typical road bicycle, the relationship between the rider's force on the pedal F_{pedal} and the force of the road F_{road} on the rear tire is $F_{\text{pedal}} = 10.5 \times F_{\text{road}}$ when the bike is in its highest gear.

To overcome the 24.5 lb of opposing force, our rider must exert a downward force on the pedal of

$$F_{\text{pedal}} = 10.5 \times 24.5 \text{ lb} = 257 \text{ lb}$$

This is a very large pedal force for a 161 lb rider. In addition to putting all of his weight on one pedal, he must also pull upward on the handlebar with a force equal to 96 lb (257 lb − 161 lb = 96 lb). Whether other gears are a better option is considered in chapter 11. Also, be careful of the wheelie, especially going uphill!

6.6 Terminal speed

In this section, we explore an interesting phenomenon that arises when objects fall under the force of gravity and simultaneously encounter significant air resistance.

> An object falling under the downward force of gravity and the upward resistive force of air resistance often attains a constant maximum speed called the *terminal speed* (or terminal velocity). The use of the word 'terminal' has nothing to do with the fact that the object might smack into the ground with unfortunate consequences. Rather, the word 'terminal' refers to the fact that, when the object initially begins to fall under the pull of gravity, it is in free fall with the acceleration of gravity. As the falling body picks up speed, the force of air resistance, proportional to the square of the speed, increases rapidly. Eventually, the upward force of air resistance equals the downward pull of the body's weight—the resultant force is zero and, in accordance with Newton's laws, the body now falls at a constant speed toward the ground. This constant speed is called the *terminal speed.*

Consider a large stone of weight 20 lb dropped out of an airplane. When it is first released, its speed is zero. Thus, there is no force of air resistance; the net force is simply the stone's weight and it, therefore, accelerates at 32.2 ft/s^2. Under this acceleration, the falling stone picks up speed and the upward force of air resistance begins to build. At some point in the descent, the force of air resistance might be 8 lb upward while the weight is still 20 lb downward. With a net force of 12 lb down, the stone still accelerates, gaining additional speed. Eventually, if the stone falls far enough, the speed increases to the point at which the upward force of air resistance equals the downward pull of gravity. At this point, the net force on the stone is zero and there is no further acceleration. For the rest of the way down, the stone travels at this constant speed—this is named the 'terminal speed'. If the stone had been dropped from a sufficiently high altitude, it can fall many miles at its constant terminal speed. An important aspect of terminal speed is that the heavier the object, the greater will be its terminal speed. In certain athletic events, such as downhill ski racing, the speeds come close to terminal for a human in a compact position. The larger skiers have the advantage in the straight downhill sections; this advantage probably tilts toward the smaller racers when it comes to making a turn.

The bigger they are, the harder they fall.

Robert Fitzsimmons, an early 20th century boxer

Since this is a book about bicycles, we wonder what would be the terminal speed of our standard cyclist if she rode out the back of a plane and maintained her cycling stance heading toward the ground. To obtain the terminal speed, we begin with expression we developed in section 5.13.

$$F_{\text{air resistance}} = 0.00640\ v^2, \qquad \text{speed in ft/s} \tag{6.6}$$

Terminal speed occurs when the downward pull of gravity on the 181 lb bike/rider system equals the upward air resistance.

$$181\ \text{lb} = (0.00640\ v^2)$$

$$v = \sqrt{\frac{181}{0.00640}} = 168\ \text{ft/s} = 115\ \text{mph}$$

The heavier an object, the greater will be its terminal speed.

6.7 Terminal speed coasting downhill on a bike

Let go thy hold when a great wheel runs down a hill.

King Lear, ii,4 William Shakespeare

Although falling out of an airplane on a bicycle is a rare event, the preceding concepts have great significance to a cyclist. In this section, we apply the concept of terminal speed to a rider coasting down a hill—the essence of the joy of the bicycle! The attainment of terminal speed does not require falling from high altitudes; it does not even require freefall through the atmosphere. A cyclist, freewheeling down a hill, will attain terminal speed when the *downward pull of the slope force is balanced by the upward resistive forces* (figure 6.3). In chapter 5 equation (5.17), we saw that the total resistive force F_{tot} on a moving cyclist consists of constant rolling and bearing resistance F_{rb} and a force of air resistance that increases with the square of the bike's speed.

$$F_{\text{tot}} = F_{\text{rb}} + k\ v^2 \tag{6.7}$$

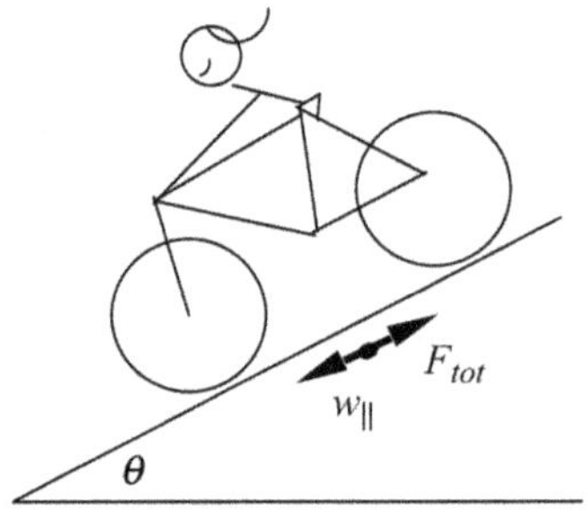

Figure 6.3. Terminal speed down a hill.

$$\text{at terminal speed: } F_{slope} = F_{tot}$$

using an expression for the slope force equation (6.4) we obtain:

$$w \times \sin\theta = F_{rb} + kv^2 \tag{6.8}$$

putting in the values for k (0.0138 when speed is in mph), a rider/bike weight of 181 lb, and rolling resistance of 1.0 lb, and bearing resistance that add to 1.006 lb, we obtain:

$$181 \text{ lb} \times \sin\theta = 1.006 + 0.0138\ v^2$$

Solving for the terminal speed with the angle as the variable:

$$v = \sqrt{\frac{181 \text{ lb} \times \sin\theta - 1.006 \text{ lb}}{0.0138\frac{\text{lb}}{v^2}}} \tag{6.9}$$

As an example, suppose the slope is a 3%, $\theta = 1.718°$ (section 6.4):

$$v = \sqrt{\frac{181 \times \sin 1.718 - 1.006}{0.0138}} = \sqrt{\frac{5.426 - 1.006}{0.0138}} = \sqrt{320} = 17.9 \text{ mph}$$

With no pedaling effort, our rider happily cruises down this hill at a brisk speed. No doubt, riding 18 mph on a level road requires a substantial effort by the rider.

Table 6.2 lists terminal speeds for our standard rider for a variety of slopes.

Table 6.2. Terminal speed versus slope.

Terminal speeds at various slopes		
Weight of bike/rider system(in lb) = 181		
Coefficient - air resistance for speed in mph = 0.0138		
Slope (%)	Angle(degrees)	Terminal speed (mph)
1	0.573	7.7
2	1.146	13.8
3	1.718	17.9
4	2.291	21.2
5	2.862	24.1
6	3.434	26.7
7	4.004	29.0
8	4.574	31.2
9	5.143	33.2
10	5.711	35.1
15	8.531	43.3
20	11.310	50.0
25	14.036	55.7
30	16.699	60.8

With a rider/bike system that is heavier, the terminal speed will be higher. If you really want to enjoy a fast downhill coast, ride a tandem. A pair of sturdy tandem riders can have a system weight of 400 lb and, since the second rider is tucked lower, the frontal area is nearly the same as for a single rider; rolling and bearing resistance will also vary little. Thus on the same 3% slope, the terminal speed for our tandem pair is:

$$v = \sqrt{\frac{400 \times \sin 1.718 - 1.006}{0.0138}} = \sqrt{\frac{12.0 - 1.006}{0.0138}} = \sqrt{796} = 28.2 \text{ mph}$$

It is certainly true that 'the bigger they are, the faster they fall!'

6.8 Personalized determination of resistive force parameters

Is there a simple way to determine the forces of resistance on my own personal bicycle?

To this point, we have done most of our calculations on a bike/rider system with standard properties. We now develop a simple technique that can be used by cyclists to determine their personalized resistive forces on their individual bicycle. The method, despite being extremely straightforward and easy to implement, allows riders to fine tune their riding stance to minimize the dominant effect of air resistance.

Table 6.2 was created assuming that we know both the rolling/bearing resistance of a bicycle and the air resistance coefficient of a bike/rider system. The resistances were introduced in chapter 5 as typical values. However, do you actually know these resistance values for your own specific bicycle/rider combination? Probably not—but the phenomena of coasting downhill to a terminal speed creates an interesting technique that allows for the personalized determination of the resistance values. *The method is very simple—find two hills of uniform slope and coast down these hills to the point of terminal speed.* We need only to record the values of the terminal speed and the angles of the two hills. An inexpensive cycle computer is adequate for speed determination; the angle of the hills may be measured in various ways; common digital protractor-levels or smart phone apps are accurate to a tenth of a degree. For purposes of calculation, it is important to use angles; slope percentages need to be converted into angles (equation (3.2)).

For two hills of different slopes, we determine the angle of the hill and the terminal coasting speed. It is important to maintain the same body position during the descent. We then apply to each hill the expression (6.8) from the previous section:

$$w \times \sin\theta = F_{rb} + kv^2 \tag{6.8}$$

Calling the hills #1 and #2 results in two expressions:

$$w \times \sin\theta_1 = F_{rb} + kv_1{}^2 \tag{6.10}$$

$$w \times \sin\theta_2 = F_{rb} + kv_2{}^2 \tag{6.11}$$

As an example, suppose on the first hill of slope 2% ($\theta_1 = 1.146°$), we reach a terminal speed of 11 mph, and the second hill of slope 5% ($\theta_2 = 2.862°$) results in a terminal speed of 25 mph.

Keeping the weight of the system (181 lb) the same, we substitute the numerical values for each slope:

$$181 \times \sin 1.146 = F_{rb} + k \times 11^2 \quad \text{or} \quad 3.62 = F_{rb} + 121\ k \tag{6.12}$$

$$181 \times \sin 2.862 = F_{rb} + k \times 25^2 \quad \text{or} \quad 9.04 = F_{rb} + 625\ k \tag{6.13}$$

These equations create the nice algebraic situation of two unknowns F_{rb} and k and two equations; our algebra teachers tell us it is straightforward to solve for the unknowns. Subtracting the left sides and the right sides (equation (6.13)—equation (6.12)) of the above expressions, we obtain

$$5.42 = 504\ k$$

For the air resistance coefficient:

$$k = \frac{5.42}{504} = 0.0108$$

substituting this value for k into equation (6.12): $3.62 = F_{rb} + 121 \times 0.0108$, we solve for the combined rolling/bearing resistance:

$$F_{rb} = 2.31\ \text{lb}$$

Perhaps this rider is in a more crouched stance that results in a smaller coefficient of air resistance than the standard rider. Other possibilities are she has a narrower frontal area or is riding at a higher altitude. However, her bike has more than twice the value for the rolling/bearing resistance. Perhaps, her machine needs air in the tires or grease in the wheel hubs!

The use of the specific rider values $k = 0.0108$ and $F_{rb} = 2.31$ lb allows for the construction of a 'personalized' table similar to table 6.2. Our cyclist now has her own personal version of equation (6.7):

$$F_{tot} = F_{rb} + k\,v^2 = 2.31 + 0.0108\,v^2$$

The reader is encouraged to go for a nice coast down two hills and develop her own values. These numbers will be extremely useful in chapter 8 when we examine the power expenditures of climbing hills.

A bit of further reflection on this chapter's last two sections reveals that a bike's terminal speed affords instant feedback on the efficiency of the riding stance. Assuming a more crouched position quickly results in a higher terminal speed. One can also test various wheels and tires for rolling resistance. Who needs a wind tunnel?

Chapter 7

Momentum–impulse

Get a bicycle. You will not regret it. If you live.

Taming the Bicycle by Mark Twain

The word 'momentum' is common in everyday conversation—one might say that a certain football team is entering the final quarter with a three-touchdown lead and, therefore, has the momentum. A sportscaster might say that the fullback's momentum carried him past the goal line. When physics concepts and terminology are incorporated into daily language, we must ensure that the word is carefully defined and applied in the physical context. Momentum derives from the Latin word for movement, and it is easy to get the concept of momentum confused with the concept of inertia.

7.1 The basic physics of momentum

Before proceeding with a precise definition of momentum, think of a situation in which you are going to catch and stop a ball. Which might you prefer?

- A golf ball traveling at 100 mph?
- A bowling ball traveling at 10 mph?

Which ball is most likely to knock you over as you try to catch it? As you think about it, the mass of the bowling ball is certainly many times that of the golf ball, yet that golf ball is traveling very fast. A dilemma? There is an answer—it comes from the physics concept of momentum. Consideration of the balls' momenta (plural of momentum) helps decide. The definition of the physical quantity *momentum* involves both the mass of the object and the velocity of the object. Both mass and velocity are of equal importance.

The *momentum* of an object is defined as the *mass of an object times its velocity*:

$$p = m\,v \tag{7.1}$$

doi:10.1088/978-1-6817-4441-4ch7

Traditionally, the letter p is the symbol for momentum. Since velocity is a vector and mass is a scalar, the product is a vector. We can calculate the momentum for the golf and bowling balls.

A golf ball has a regulation weight of 1.62 oz; this converts into a mass of 0.00316 slugs. The ball's velocity of 100 mph is 147 ft/s. Hence, the momentum of the golf ball is:

$$p = mv = 0.00316 \text{ slugs} \times 147 \text{ ft/s} = 0.465 \frac{\text{slugs} - \text{ft}}{\text{s}}$$

There is no special name for the units of momentum.

We do the same calculation for the bowling ball. Suppose the bowling ball has a weight of 16 lb or 0.497 slugs. The ball's velocity of 10 mph is 14.7 ft/s. The momentum of the bowling ball is:

$$p = mv = 0.497 \text{ slugs} \times 14.7 \text{ ft/s} = 7.31 \frac{\text{slugs} - \text{ft}}{\text{s}}$$

The traveling bowling ball has a much greater momentum and, as we will see, it will require a much greater force to stop.

7.2 Momentum and Newton's Second Law

Momentum may be envisioned as evolving from Newton's Second Law:

$$F = ma$$

Since acceleration is change in velocity divided by change in time, we write:

$$a = \frac{\Delta v}{\Delta t}$$

The Second Law becomes:

$$F = ma = \frac{m\Delta v}{\Delta t}$$

Since mv is p *the momentum,* $m\Delta v$ is Δp, *the change in momentum*

$$F = \frac{\Delta p}{\Delta t} \tag{7.2}$$

We have an alternative form of Newton's Second Law—*net external forces produce changes in momentum.*

7.3 Impulse

Beginning with this new form of Newton's law (equation (7.2)) and doing a simple rearrangement, we obtain:

$$F \times \Delta t = \Delta p \tag{7.3}$$

The quantity $F \times \Delta t$ (force × change in time) is called the 'impulse'
In words:

$$\text{impulse} = \text{change in momentum}$$

The change in momentum will be the final momentum p_f minus the initial momentum p_i, thus:

$$F \times \Delta t = p_f - p_i \tag{7.4}$$

Impulse is an extremely useful physical concept that assists in understanding the behavior of moving objects. Consider a familiar situation such as a catcher catching a fastball. When the moving ball is brought to rest, the final momentum p_f of the ball in the mitt is zero; hence, the change in momentum equals the momentum of the ball just before it reaches home plate p_i:

$$\Delta p = p_f - p_i = 0 - p_i = -p_i \tag{7.5}$$

The impulse momentum expression now becomes:

$$F \times \Delta t = \Delta p = -p_i \tag{7.6}$$

The catcher has no control over the moving ball's initial momentum and the change in momentum (unless he lets it somehow slip past his mitt). The momentum change is determined entirely by the speed of the ball just before it enters his glove. However, the catcher does have control over how badly the ball stings his hand. When catching the ball, he wisely moves his glove backward—thus extending the time of contact. The impulse, the product of force and Δt, will be the same regardless of how the catch is made; by moving the glove backward (making Δt as large as possible), the catcher makes the force as small as possible.

A boxer who 'rolls with the punch' is performing an identical action; he is extending the time of contact of his opponent's glove, thereby reducing the force exerted by the glove. The opposite situation occurs when a boxer 'steps into a punch'. Stepping into the punch makes Δt small and consequently maximizes the force on the jaw.

7.4 Momentum and impulse aspects of bicycle accidents

Why wear a helmet? What is the physics behind how it actually protects your head?

Momentum and impulse allow us to appreciate the wisdom of wearing a helmet. Consider the moving mass to be the rider's head; suppose it is traveling at the speed of the bike when the rider hits an obstacle and his head comes to rest. The role of the helmet is to extend the time of the collision; by making the time for the momentum change as large as possible, the force is minimized.

How might Δt be maximized? Primarily it occurs by the deformation of the outside shell and subsequent compression of the foam padding. As the protective materials 'give', they extend the time of contact during which the rider's head is being brought to rest. The larger the Δt, the smaller will be the force on the head. The mechanics is similar to the catcher and ball.

These are the same principles used in automobile safety equipment—the seat belts are designed to stretch and the air bags to compress. These actions extend the time of contact and minimize the forces on the car's occupants. Would it make sense to replace the flexible fabric seat belts with heavy steel chains that will not readily stretch? What if we lined the bike helmet with a concrete like material molded to fit exactly the shape of the rider's head?

Suppose we equipped the inside of the bike helmet with highly elastic springs that would compress and slow down the moving head. Would this be a good idea instead of compressible padding? No, because the head would bounce backward with a final speed equal to its initial speed.

If you can keep your head when all about you
Are losing theirs and blaming it on you

If by Rudyard Kipling

Why is a fall that results in a long slide usually safer than a fall that ends in a quick stop on the ground?

The most serious damage from a fall occurs from large impact forces. With the long sliding fall, the time (Δt)for the rider to come to rest is large; the longer time results in a smaller force. Although the force is small, it is usually absorbed by the rider's skin that is not especially good at handling forces. Nasty road burns usually result; not pleasant, but probably better than broken bones caused by quick hard falls onto arms, shoulders, ribs.

Chapter 8

Work–energy–power

It's true hard work never killed anybody, but I figure, why take a chance?
President Ronald Reagan

Working hard climbing that hill, burning enormous energy in a long ride, powerful sprints at the finish of a race—are phrases familiar to cyclists. However, the words *work*, *energy*, *power* have so many different meanings that vary with the context. This chapter is an examination of the physical concept of work and its association with energy and power. As previously seen, words that are part of our everyday vocabulary must be carefully defined within the physical context.

8.1 Work

We begin with a word that evokes a variety of emotional responses when used in daily conversation. The word is *work*. How often have we said or heard statements such as:

- Studying physics involves a lot of work
- I have a lot of expenses and need to find some work that pays me at least \$20/hour
- You need to work through the solution to this problem
- If you work out, it will help with your weight loss program
- I am so tired after coming home from work I need a three hour nap

These uses of the word 'work' are quite different from the concept of work as we define and use in physics. In the world of physics, the following are not necessarily examples of work; in fact, they may involve no work whatsoever:

- actions you are paid to do
- actions that make you tired
- actions that are hard to do
- actions that you do to smooth out problems in relationships

doi:10.1088/978-1-6817-4441-4ch8

If the above do not define work in a physical sense, what is work's physical meaning? The definition of work is based on the exertion of a force and movement:

work = force exerted times the distance moved in the direction of the force

In letters:

$$W = F \times d \tag{8.1}$$

Where: W is work done, F is force doing the work, and d is the distance moved in the direction of the force.

Consider a simple example—we push a 20 lb box across a 15 ft wide room with a horizontal force of 10 lb (figure 8.1A).

The work done by the force is:

$$W = F \times d = 10 \text{ lb} \times 15 \text{ ft} = 150 \text{ ft-lb}$$

Although the force and the distance moved are both vectors, their product 'work' is a scalar quantity—it has no direction; work has magnitude only. In this example, the magnitude is 150 ft-lb. In U.S. Customary Units, the unit of work is the **ft-lb**. In the metric system, where force is in newtons and distance is in meters, the unit of work is the **N-m** (newton-meter), also called the joule.

The weight of the box did not factor into the work done by the pushing force. The distance moved is in the horizontal direction, whereas the weight of the box is a vertical force. Instead of pushing the box along the floor, we lift the box to waist level, 3 ft off the floor, and proceed to carry it 15 ft across the room. Is there any difference in work done as compared to sliding it with the 10 lb push? There will be work done in the lifting process—to raise it off the ground; we must counter its weight by exerting a force of 20 lb in the vertical direction. Lifting the box to a height of 3 ft will require the work:

$$W = F \times d = 20 \text{ lb} \times 3 \text{ ft} = 60 \text{ ft-lb}$$

Of course, if we are holding the box off the ground, we must continue to exert this 20 lb vertical force. If we now walk 15 ft across the room, is there an additional work of

$$W = F \times d = 20 \text{ lb} \times 15 \text{ ft} = 300 \text{ ft-lb?}$$

No, holding the box (even for hours) or carrying it across the room may involve no work in the physics sense. The reason is that the exerted force is vertical and the

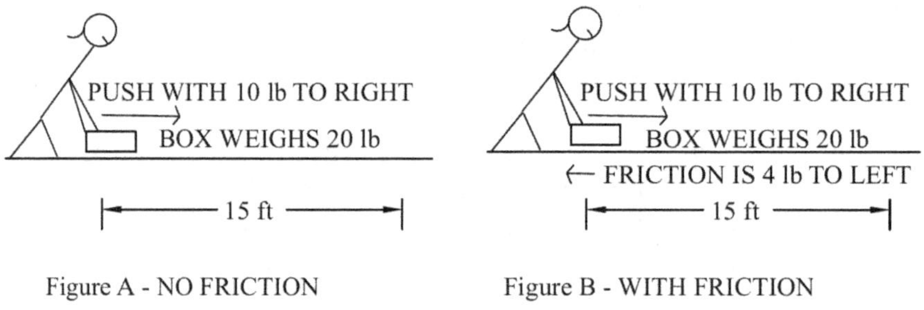

Figure 8.1. (A,B) Push box across room.

movement across the room is horizontal. After the initial lift, there is no further movement in the vertical direction. The calculation of work done in carrying the box across the room looks like:

$$W = F \times d = 20\text{ lb} \times 0\text{ ft} = 0\text{ ft-lb}$$

Carrying the box might make you tired but the horizontal motion involves little or no work. What if you carry it 15 000 ft? There is still no work! No doubt, your arms will ache and you will be very tired—but, probably, there is no additional work beyond that involved in the initial lift. Why the equivocal term 'probably' in the previous sentence? We are assuming a scenario of walking at a constant velocity across the room. Yes, there would be a horizontal force required to accelerate the box from rest to some small horizontal velocity. If the acceleration is gradual, this force of acceleration can be very small. Once you have begun walking with the box, you would likely move at a steady velocity. All of your efforts are directed toward holding the box upward against the force of gravity.

What if we carry the 20 lb box up a 16 ft high stairway? If we do not impart significant speed to the box and merely carry the box at some slow steady rate, the work done by the 20 lb vertical force through a vertical distance of 16 ft will be:

$$W = F \times d = 20\text{ lb} \times 16\text{ ft} = 320\text{ ft-lb}$$

Naturally, the stairway involves some horizontal movement, but only vertical displacements result in work done by the vertical force. What of other methods of vertical travel? A shallow ramp? A steep ramp? A circular ramp? Shimmy up a rope while hanging onto the box? Since they all involve the same vertical rise of 16 ft, the work done will be the same in each case, 320 ft-lb. Is it not harder to shimmy up the rope than walk up the ramp? Yes—but the difficulty of performing an action is a not a factor in the amount of work performed.

What if you sit at a desk all day, maybe talking on the phone; your company pays you quite well for your knowledge—is this considered work? Not in the physics sense! Unless you are exerting forces in the direction moved, you are not doing physical work. Perhaps, talking on the phone involves a small of work as you lift the receiver to your ear.

I like work, it fascinates me. I can sit and look at it for hours.

Jerome K Jerome (1859–1927)

8.2 Kinetic energy

Why such a restrictive definition for the term 'work'? This precise definition of work gives meaning to the very important concepts of energy and power. To gain a better understanding of the consequences of performing work, let us return to our original example of pushing a 20 lb box with a 10 lb force across the 15 ft room. To keep things simple, suppose the contact between the box and the floor is frictionless; the floor is waxed or the box is on wheels. From Newton's Second Law, we can determine the acceleration of the box. The 20 lb box will have a mass of 0.621 slugs, and the acceleration caused by the 10 lb push is:

$$a = \frac{F}{m} = \frac{10\ \text{lb}}{0.621\ \text{slug}} = 16.1\ \text{ft/s}^2$$

The 10 lb force is causing an increase in the box's speed. In this chapter, we develop a new way of thinking about the actions—we have seen that exertion of a force, through a distance in the direction of the force, results in the performance of a physical quantity called work. The box gains motion; we say it possesses *kinetic energy.*

Kinetic energy is the energy of a body associated with its motion.

We can obtain an expression that allows the calculation of the kinetic energy of a moving mass. Assume the box is initially at rest and a constant force F accelerates the box from zero to some final speed v. The work done is

$W = F \times d$ and writing the force as $F = ma$ we obtain:

$$W = m \times a \times d \tag{8.2}$$

We can make use of an equation (4.4) from chapter 4:

$$d = \frac{v_{\text{ave}}}{2} = \frac{v_f + v_i}{2}t = \frac{v}{2}t$$

and the acceleration of the object is written:

$$a = \frac{v}{t}$$

Making these substitutions into (8.2), the expression for work becomes:

$$W = m\left(\frac{v}{t}\right)\left(\frac{v}{2}t\right) = \frac{mv^2}{2}$$

The work done on the box has been converted into kinetic energy KE of the box. We have a general expression for the kinetic energy of a moving object:

$$KE = \frac{mv^2}{2} \tag{8.3}$$

In U.S. Customary Units, the units of kinetic energy are the same as those of work: *foot-pounds* (ft-lb). In the metric system, the units will be the *newton-meter* (N-m), or the *joule.* Just as work is a scalar quantity, so too is kinetic energy—the kinetic energy of a moving object has no direction, only a magnitude. In addition to kinetic energy, we will see other forms of energy and they are all scalar quantities. The scalar characteristic of energy is an important reason for its popularity in analyzing physics situations. Although the work creates the kinetic energy of the box, and work and kinetic energy have the same units of measurement, *they are very different concepts. The work is done by the force of the person; the kinetic energy is gained by the box.* One of our principals (the person) does the work and the other principal (the box) is the beneficiary.

As a simple example, consider the case from the previous section in which a 10 lb force was exerted on a 20 lb (0.621 slugs) box (figure 8.1A). The work done was 150 ft-lb; with no frictional losses, the kinetic energy would also be 150 ft-lb. The speed of the box may be found from equation (8.3):

$$150 \text{ ft-lb} = \frac{0.621\, v^2}{2}, \qquad \text{thus} \qquad v = 22 \text{ ft/s}$$

Kinetic energy is a type of energy called *mechanical energy*; there are other forms of mechanical energy. Especially important is gravitational potential energy

Does the work done always lead to a gain in energy? Not always! In our examples thus far, the force and the distance moved have been in the same direction and the work done is a positive quantity. Positive work results in gains in energy. A bit of reflection will reveal that it is possible to exert a force opposite to a direction of motion. For instance, suppose the box is already traveling to the right when we encounter it. We might exert a force to the left in an attempt to stop the box. In this situation, the direction of the box's motion is opposite to the direction of the force. The work done is a negative quantity that results in a loss of kinetic energy for the box.

8.3 Frictional effects

In figure 8.1B, we see another example in which a box is pushed along on a floor; in this situation, there is friction between the bottom of box and the floor. Friction is opposite the direction of travel; therefore, the work done by friction force is negative. Assume the coefficient of sliding friction is 0.2. Since the floor exerts a vertical force N equal to the weight of the box, the force of sliding friction is:

$$f = \mu \times N = 0.2 \times 20 \text{ lb} = 4 \text{ lb}$$

The force of friction will do a negative work:

$$W = f \times d = -4 \text{ lb} \times 15 \text{ ft} = -60 \text{ ft-lb}$$

As before, the work done by the girl pushing will be 150 ft-lb. This negative work results in less kinetic energy for the box when it gets to the right side of the room. The relationship between the change in kinetic energy of the box and the positive and negative work done on the box is as follows:

$$\text{gain in } KE = \text{positive work} - \text{negative work} \tag{8.4}$$

$$\frac{1}{2} m\, v^2 = F \times d - f \times d \tag{8.5}$$

Putting in numbers: ½(0.621 slugs) v^2 = (10 lb × 15 ft) – (4 lb × 15 ft) and solving for the speed: v = 17 ft/s.

The final speed is less when compared with the result of 22 ft/s (section 8.3) when there was no friction; frictional losses cost kinetic energy and speed. Unfortunately, the energy lost to friction goes into heating the surfaces of the box bottom and the floor; there is no way to recover these losses to heat. The force of friction is an example of a *non-conservative* force. The opposite of a non-conservative force is a *conservative* force. With a conservative force, there are no losses of mechanical energy—only transfers from one type of mechanical energy to another. The most common example of a conservative force is the force of gravity. The kinetic energy of a bike that is taken away on a ride uphill is returned to the system on the downhill trip. Sadly, air resistance and other resistive forces are non-conservative and result in loss of mechanical energy.

8.4 Gravitational potential energy

The easiest demonstration of conservative transfers of mechanical energy is to toss a ball up into the air. As the ball is moving up, the force of gravity points down and the work done by the gravitational force is *negative*; the negative work done by gravity causes the ball *to lose* its kinetic energy. At some point, the ball has lost all kinetic energy and has a speed of zero; the ball now begins to fall back to earth and moves in the same direction as the force of gravity. For the falling ball, the work done by gravity is *positive* and the ball *gains* kinetic energy.

Think further about the ball when it is at the top of its flight—it has momentarily stopped moving and therefore has no kinetic energy—the negative work done by gravity has 'taken away' the kinetic energy. Nevertheless, at the top of its flight, the ball does have something; in the next instant, it will begin to fall and regain kinetic energy. Even if the ball is stuck in the top of a tree, it has 'something' when it is up high. The ball possesses energy of position; this energy of position is called *potential energy*. The energy is due to the ball being subject to a gravitational force; the ball has *gravitational potential energy*.

Does an object first have to start with kinetic energy (the tossed ball) in order to later possess gravitational potential energy? No—another simple, common event—suppose you bend over and lift an object off the floor. Maybe it is a young child—you have no desire to impart kinetic energy to the kid. You slowly lift the child and hold him at waist level. If the child weighs 20 lb and you lift him a distance 4 ft off the floor, your work is:

$$W = F \times d = 20 \text{ lb} \times 4 \text{ ft} = 80 \text{ ft-lb}$$

As you stand holding the child in the air, what happened to the work you performed? Your work has been transformed; the toddler now has potential energy as a result of his high position.

Gravitational potential energy is the energy possessed by an object by virtue of its position being subject to the gravitational force.

Notice the terminology—the gravitational potential energy is possessed by the object—you did the work—but now the object has the potential energy. You could put the body on a shelf and walk away; the object still has the energy. Please do not put the kid on a shelf and walk away!

To get an expression for the gravitational potential energy, let w be the weight of the body and Δh be the vertical distance the object is lifted

work done = gain in gravitational potential energy

It is common to use the letters PE to symbolize potential energy; a subscript indicates gravitational potential energy. It is also common to write the weight as the mass times the acceleration of gravity; thus, we get:

$$PE_g = w \times \Delta h = m \times g \times \Delta h \tag{8.6}$$

There is an aspect of gravitational potential energy that sometimes causes a little confusion. We discussed lifting the child to a height 4 ft off the floor, but what if this were done in a second story room—maybe 12 ft off the ground? A question arises as to which height to use: 4 ft, 12 ft or even 16 ft? Keep it simple—just use the height the object will drop.

8.5 Conservation of energy

We have seen examples in which kinetic energy is converted into gravitational potential energy and, once an object has potential energy, it can fall and regain its kinetic energy. When the objects falls, its gain in kinetic energy comes at the loss of potential energy. These transformations between various types of energy illustrate one of the most profound and significant principles of the physical world—the *principle of conservation of energy.*

Principle of Conservation of Energy

The total amount of energy of a system cannot be created or destroyed, rather only converted from one form of energy to another.

We write the conversion between the two forms of mechanical energy as:

changes (gains or losses) in potential energy = changes (losses or gains) in kinetic energy

Another way to describe the energy conservation of an event as:

$$(\text{kinetic energy} + \text{potential energy})_{\text{initial}} = (\text{kinetic energy} + \text{potential energy})_{\text{final}}$$

What happens if a 20 lb weight, initially at rest, falls 5 ft to the floor? Consider the floor to be the zero point of potential energy and ignore frictional losses to air resistance. The principle of conservation of energy tells us:

$$(KE + PE)_{\text{initial}} = (KE + PE)_{\text{final}} \tag{8.7}$$

Since the initial KE and the final PE are zero, we write:

$$PE_{\text{initial}} = KE_{\text{final}} \tag{8.8}$$

substituting $KE = 1/2\ mv^2$ and $PE = mgh$:

$$(mgh)_{\text{initial}} = \left(\frac{1}{2}mv^2\right)_{\text{final}} \tag{8.9}$$

Solving for the speed at which the object hits the ground:

$$v = \sqrt{2gh} \tag{8.10}$$

Notice the speed does not depend on mass of object (in absence of air resistance). For an object dropped from a height h of 5 ft:

$$v = \sqrt{2 \times 32.2\ \text{ft/s}^2 \times 5\ \text{ft}} = 17.9\ \text{ft/s} = 12.2\ \text{mph}$$

This result, obtained from energy principles, is the same if obtained using Newton's Second Law.

What of a bike rider who flies down a hill 500 ft high? Can we say that she reaches the bottom of the hill at a speed:

$$v = \sqrt{2 \times\ 32.2\ \text{ft/s}^2 \times 500\ \text{ft}} = 179\ \text{ft/s} = 122\ \text{mph}$$

No—in chapter 5, we saw the forces of resistance (especially air) are very significant on a moving cyclist. It cannot be ignored; the simple expression for speed at the bottom does not apply. One would have to subtract the energy losses to the negative work done by the resistive forces. The speed dependence of air resistance requires a calculus based analysis.

Before leaving this section, there is one additional consideration—we have been looking at the conversion of gravitational potential energy into kinetic energy. Conservation of energy is reversible; a body projected upward has its kinetic energy converted to potential energy; the height of the rise may be then calculated.

$$\begin{aligned} KE_{\text{initial}} &= PE_{\text{final}} \\ \frac{1}{2}m\ v^2 &= mgh \\ h &= \frac{v^2}{2g} \end{aligned} \tag{8.11}$$

8.6 Energy conversion between kinetic and potential on the bicycle

We can consider the application of energy concepts to a bicycle ride of our typical rider of weight 161 lb on a 20 lb bicycle (system mass of 5.62 slugs). At a speed of 20 mph (29.3 ft/s) the kinetic energy is:

$$KE = \frac{1}{2}m\ v^2 = \frac{1}{2}\ 5.62\ \text{slugs} \times (29.3\ \text{ft/s})^2 = 2412\ \text{ft-lb}$$

Once the bicycle is at the desired speed and kinetic energy, where does this energy go if you stop pedaling? The kinetic energy of the coasting bike is converted into frictional losses against air, rolling, and bearing resistance (chapter 5) and, if you coast up a hill, into gravitational potential energy. Suppose the losses to resistive forces are ignored, how high up the hill could you travel from the speed of 20 mph? We use the expression from the last section (equation (8.11)).

$$h = \frac{v^2}{2g} = \frac{(29.3\ \text{ft/s})^2}{2 \times 32.2\ \text{ft/s}^2} = 13.3\ \text{ft}$$

Not very high and the situation is even worse when the resistive forces are considered.

8.7 Power—the basic physics

Once again, we encounter a concept in physics that is also common in our daily vocabulary. It is important to carefully and precisely define the term before it is used in physics. The concept is *power*. You might hear someone say:

- We lost power during the ice storm
- That car has a powerful engine
- Superman is more powerful than a locomotive
- The fullback powered his way through the defensive line
- The hitter is a powerful slugger
- The lobbyists have all the power in Washington

These uses of the word *power* imply a source of energy or the ability to exert large forces. These differ from the phenomenon of *power* in the physical word.

power is the rate of doing work

or

power is the energy gained or lost per time

Before exploring these precise definitions, consider a bizarre example. Suppose we have 500 lb of sand and need to move it from the ground to the top of a roof 20 ft high. Do not ask why—the boss says to move the sand—just start moving the sand. Perhaps there are three options: you could personally move the sand, or you could ask Superman to move the sand, or maybe you have a trained ant that could move the sand; have you ever noticed ants love to move and make little piles of sand (figure 8.2)?

The work done on the 500 lb of sand in moving it a vertical distance of 20 ft is:

$$W = F \times d = 500\ \text{lb} \times 20\ \text{ft} = 10{,}000\ \text{ft-lb}$$

It does not matter who actually moves the sand—you, Superman, or the ant. Nevertheless, there is a big difference in how the three individuals get the job done; the difference will be the time required:

- the person—would probably make many trips. Suppose you carry 50 lb at a time—that will be 10 trips. Let us give three minutes per round trip. Total time for the human is 1 800 s.
- Superman—is described as 'faster than a speeding bullet'. Surely, he could scoop all 500 lb of the sand into his cape and, since he is 'able to leap tall

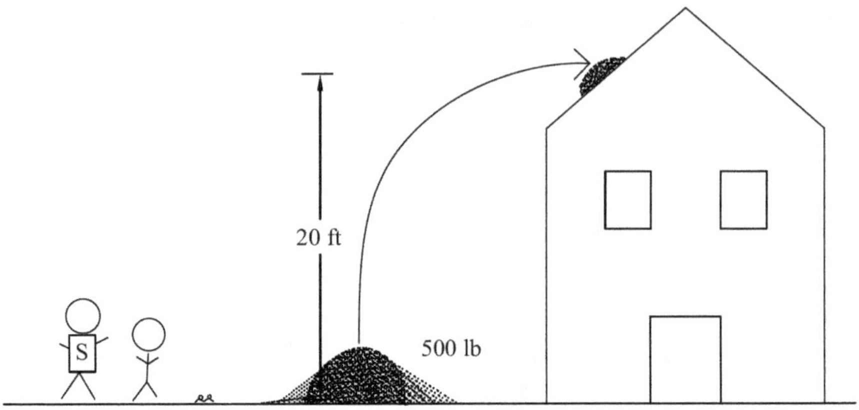

Figure 8.2. Sand on roof of house.

buildings with a single bound', he could probably just jump to the rooftop in a quarter of a second. Total time for Superman is 0.25 s.
- the ant—let's just take a wild guess and suppose it takes him about 100 years; this is about 3 billion seconds

The three characters—the human, Superman, and the ant—can all do the *same work*: 10 000 ft-lb. Where they differ is how long it takes to do this work. The concept of *power*, the rate of doing work, is the true distinction between a human, Superman, and an ant. Power P is the amount of work done per unit of time, mathematically:

$$\text{power} = \frac{\text{work}}{\text{time}} = \frac{W}{t} \tag{8.12}$$

We calculate the powers of the human, Superman, and the ant by dividing work done by their times (table 8.1).

Table 8.1. Various power levels to move sand.

	work (ft-lb)	time (seconds)	power (ft-lb/s)	power (hp)	power (watts)
human	10 000	1 800	5.56	0.010	7.54
Superman	10 000	0.25	40 000	72.8	54 200
ant	10 000	3×10^9	3.33×10^{-6}	6.09×10^{-9}	4.52×10^{-6}

Since ft-lb is the unit of work and energy in U.S. Customary units, ft-lb/s is the unit of power in these units; alternate units of power are the *horsepower* and, in the metric system, the *watt*.

These units of power may be obtained through the following conversions:

$$1 \text{ horsepower (hp)} = 550 \text{ ft-lb/s}, \quad 1 \text{ ft-lb/s} = 0.00182 \text{ hp}$$
$$1 \text{watt} = 0.737 \text{ ft-lb/s} \qquad 1 \text{ ft-lb/s} = 1.356 \text{ watt}$$
$$1 \text{ hp} = 746 \text{ watts} \qquad 1 \text{ watt} = 0.00134 \text{ hp}$$

Another unit of power is the number of calories per hour or calories per day. As we will see in chapter 9, heat is another form of energy and the calorie is a unit of heat; thus, calorie/time is a unit of power. While the word spelled 'c-a-l-o-r-i-e' may sound familiar, it is not exactly the same as when we discuss going on a diet and losing weight or gaining weight by eating rich, fattening foods. In this context, the word is spelled 'C-a-l-o-r-i-e'—capitalization matters! *A Calorie is equal to 1000 calories.* Normally, the Calorie is used to denote the energy content of a food; look on the nutritional label of a packaged food. That serving of potato chips that contains 450 Calories actually contains 450 000 calories!

In U.S. Customary units, a calorie of heat is equivalent to 3.09 ft-lb. A calorie of heat is equivalent in metric units to 4.184 joules of energy.

You often see on a food package's nutritional label a reference to a 'normal' dietary intake of 2000 Calories per day, a measure of power. It might be useful to examine the conversion into watts:

$$\frac{2000\text{ Calories}}{\text{day}} \times \frac{1000\text{ calories}}{\text{Calorie}} \times \frac{4.184\text{ Joules}}{\text{calorie}} \times \frac{1\text{ day}}{86\,400\text{ s}} = 96.9\,\frac{\text{Joules}}{\text{s}} = 96.9\text{ watts}$$

Therefore, if you go about your day subsisting on 2000 Calories, you are slightly less powerful than a 100 watt light bulb! The 2000 Calorie/day power output corresponds to energy usage in normal everyday activities—walking around, getting up and down, climbing steps, thinking, eating, digesting, etc. A very active lifestyle will result in a higher power level. When considering a form of exercise, it is common to look at the power output in terms of watts or Calories/minute or Calories/hour.

While the power numbers, in table 8.1, for Superman are truly impressive, and we really did not expect much from the ant; what can be said of the human? A power output of 7.54 watts is nothing to brag about. Was our human really trying that hard? A point to keep in mind is that we are only calculating for the work done on the sand. Our human must also perform work to get the body's weight up to the roof level.

Each trip to the roof involves the weight of the sand and the weight of the person. Suppose the person weighs 161 lb, for one trip the work is:

$$(161\text{ lb} + 50\text{ lb}) \times 20\text{ ft} = 4\,220\text{ ft-lb}$$

for ten trips, the total work done is 42 200 ft-lb. In addition, by moving faster, maybe the job can be accomplished in 600 s. The power will be

$$P = \frac{42\,200\text{ ft-lb}}{600\text{ s}} = 70.3\,\frac{\text{ft-lb}}{\text{s}} = 0.128\text{ hp} = 95\text{ watt}$$

The above power output is the rate of doing mechanical work. However, the human body operates at a low efficiency of about 20%. Thus, the total body output will be five times as large.

Why is it easier to push a bike up a steep hill rather than ride it up the hill?

It is a question of power: whether you ride or walk the bike up the hill, you travel the same vertical distance. The gain in potential energy is the same; the work done is the same. The situation is similar to the ant, human, and superman carrying the sand to the roof of the house; the expenditure of power depends on the time taken. When you walk the bike, you can go as slow as you wish, one or two miles per hour, perhaps even slower. However, while riding a bicycle it is difficult to maintain balance at very low speeds. Walking the bike up the hill will take longer; hence, the power level is reduced.

8.8 Power and kinetic energy

In the last section, our discussion of power looked at work that resulted in a gain of potential energy. The performance of work can also lead to gains in kinetic energy or frictional heating. Sometimes, work results in an energy gain of all forms of energy—potential energy, kinetic energy, and frictional heating. What type of human power generation is associated with kinetic energy? How about sprinting a 100 yard dash in 10 s?

A speed of 100 yards in ten seconds works out to an average speed of:

$$v = \frac{300 \text{ ft}}{10 \text{ s}} = 30 \text{ ft/s} = 20.5 \text{ mph}$$

If the runner starts from rest, can we write the power as:

$$\text{power} = \frac{\text{work done}}{\text{time}} = \frac{\text{gain in kinetic energy}}{\text{time}}$$

$$\text{power} = \frac{\frac{1}{2} m v^2}{t} \tag{8.13}$$

Assume the sprinter reaches her top speed quickly and is able to maintain this speed for most of the race. Her top speed will be close to her average. This top speed is attained quickly; we will use two seconds. If her mass is five slugs, the mechanical power for the sprint is:

$$\text{power} = \frac{\frac{1}{2} m v^2}{t} = \frac{\frac{1}{2} \times (5 \text{ slugs}) \times (30 \text{ ft/s})^2}{2 \text{ s}} 1125 \frac{\text{ft-lb}}{\text{s}} = 2.05 \text{ hp}$$

The actual power expenditure will be even higher due to losses, primarily air resistance, as the runner approaches speeds near 10 mph (chapter 5). Again, the total body power output will be much larger due to the low efficiency of the body.

An important additional insight to the power requirements for a moving body can be obtained by returning to the definition of power:

$$\text{power} = \frac{\text{work}}{\text{time}} = \frac{Fd}{t} = F \times \left(\frac{d}{t}\right)$$

We have grouped the distance d and time t into the parenthesis; this ratio is simply the speed v.

Thus:

$$P = F \times v \qquad \text{in words:} \qquad \text{power} = \text{force} \times \text{speed} \tag{8.14}$$

This expression gives a method to determine the power output in a variety of situations. It is especially useful in determining the power output required to keep objects moving at constant speeds.

As a simple illustration, think of lifting a 50 lb box a height of 4 ft off the ground in a time of 1.0 s. Lifting the box at a steady speed of 4 ft/s and exerting only enough force to overcome the box's weight, there will be no gain in kinetic energy. The power can be written:

$$P = F\ v = 50\ \text{lb}\ \times\ 4\ \text{ft/s} = 200\ \frac{\text{ft-lb}}{\text{s}} = 0.364\ \text{hp} = 271\ \text{watts}$$

(The same result for power is obtained if work is calculated as 50 lb × 4 ft= 200 ft-lb divided by the time of 1.0 s.)

8.9 Power output to overcome resistive forces on a bike

Chapter 5 examined the forces of resistance on a bicycle; table 5.2 listed the resistive forces F_{tot} on a moving bicycle on a level road. Using the speed and the total resistive force, we apply the expression:

$$P = F_{tot} \times v \tag{8.15}$$

From chapter 5 the expression for total resistive force is:

$$F_{tot} = F_{rb} + F_{air} = F_{rb} + k\ v^2 \tag{8.16}$$

- F_{tot} is the total resistive force on a moving cyclist
- F_{rb} is a constant value of rolling resistance and bearing resistance:
- F_{air} is the air resistance term proportional to the square of the speed, $F_{air} = kv^2$

Using the typical values from chapter 5:

$$F_{rb} = 1.006\ \text{lb} \quad \text{and} \quad k = 0.00640, \quad v \text{ in ft/s}$$

$$F_{tot} = F_{rb} + F_{air} = F_{rb} + k\ v^2 = 1.006 + 0.00640\ v^2 \tag{8.17}$$

Combining equation (8.17) with equation (8.15) we obtain a very interesting (and disheartening) equation:

$$P = 1.006\ v + 0.00640\ v^3 \tag{8.18}$$

The demoralizing aspect is the second (air resistance) term—the power goes as the cube of the speed. If you wish to double your riding speed, it is necessary to generate eight times the power ($2^3 = 8$)!

We have used the 'k' parameter appropriate for speed in ft/s ($k = 0.00640$) rather than for speed in mph ($k = 0.0138$). This is necessary for the power calculation. Another issue in comparing the various tables of numeric calculations is minor variations (usually second decimal places) in tabulated data. These variations occur from calculation rounding.

In table 8.2, equation (8.18) is used to obtain the power to overcome these forces for a range of speeds. Included are columns that show the power in various units such as ft-lb/s, watts, horsepower (hp) and Calories per hour.

Table 8.2 comes with a major caveat—it shows the *mechanical power* needed to overcome the resistive forces. *It does not accurately portray the total power that must be generated by the cyclist's body.* The muscular efficiency of the rider's body is considered in the next section.

Table 8.2. Mechanical power to overcome air + rolling + bearing resistance.

constant = 0.006400 speed in ft/s, area = 5.382 sq ft
rolling resistance of tires = 1.0 lb
bearing resistance = 0.00628 lb

Speed		Resistive force				Total Mechanical power output			
mph	ft/s	air lb	rolling lb	bearing lb	total lb	ft-lb/s	watts	hp	Calories/hour
2	2.93	0.06	1.0	0.00628	1.06	3	4.2	0.01	5
4	5.87	0.22	1.0	0.00628	1.23	7	9.8	0.01	11
6	8.80	0.50	1.0	0.00628	1.50	13	17.9	0.02	21
8	11.7	0.88	1.0	0.00628	1.89	22	30.0	0.04	35
10	14.7	1.38	1.0	0.00628	2.38	35	47.4	0.06	55
12	17.6	1.98	1.0	0.00628	2.99	53	71.4	0.10	83
14	20.5	2.70	1.0	0.00628	3.70	76	103	0.14	120
16	23.5	3.52	1.0	0.00628	4.53	106	144	0.19	168
18	26.4	4.46	1.0	0.00628	5.47	144	196	0.26	229
20	29.3	5.51	1.0	0.00628	6.51	191	259	0.35	303
22	32.3	6.66	1.0	0.00628	7.67	247	336	0.45	392
24	35.2	7.93	1.0	0.00628	8.94	315	427	0.57	498
26	38.1	9.31	1.0	0.00628	10.3	393	534	0.72	623
28	41.1	10.79	1.0	0.00628	11.8	485	657	0.88	767
30	44.0	12.39	1.0	0.00628	13.4	589	800	1.07	933
32	46.9	14.10	1.0	0.00628	15.1	709	962	1.29	1122
34	49.9	15.91	1.0	0.00628	16.9	844	1145	1.53	1336
36	52.8	17.84	1.0	0.00628	18.8	995	1350	1.81	1576
38	55.7	19.88	1.0	0.00628	20.9	1164	1579	2.12	1843
40	58.7	22.03	1.0	0.00628	23.0	1351	1834	2.46	2140

8.10 Efficiency considerations in muscular effort

Table 8.2 summarizes the mechanical power output that the rider must expend to counter the resistive forces. However, no mechanical process is 100% efficient. In a machine such as the bicycle, there are frictional losses in the chain–gear system. Evidence for such losses is seen in the wear that occurs in the chain and gear teeth. Fortunately, these mechanical losses on a well maintained bike are amazingly small. Modern derailleur systems have an overall mechanical efficiency of nearly 95% [1]. This means that 95% of the mechanical power applied to the front chainring is delivered to the rear wheel of the bicycle; only 5% is lost to chain/gear friction. In principle, in order to account for the drive system's mechanical losses, the table 8.2 power values should be adjusted to about 5% higher. We are not going to make this small adjustment because there is another mechanical process that operates at a much lower level of efficiency. The inefficient process is in the human musculoskeletal system. At a given level of mechanical output, the body generates a high level of heat. This heat may be considered as wasted mechanical effort.

The efficiency of muscular effort depends on a variety of factors—fitness levels, training methods, pedaling rates, etc. Efficiency studies depend on careful measurements of factors such as oxygen consumption. Published values of muscular efficiency show numbers in the range of 20% [2]; a study in 2007 reveals that the cycling efficiency can vary by a factor of two fold [3].

If we adopt the 20% muscular efficiency, the overall muscular effort is a quintupling of the mechanical output. During periods of muscular exertion, the body generates a 20% mechanical output and the other 80% of the energy goes into heat. The right side of table 8.3 shows the total body power output as five times the mechanical output.

At this point in our discourse, a reasonable question might be raised as to the real-world accuracy of a table of numbers such as table 8.3. The power levels in this table are 'theoretical calculations' based on the model for resistive forces discussed in chapter 5.

A source of comparison for the calculated power levels is *The Compendium of Physical Activities* [4]. In the Compendium article, hundreds of physical activities are assigned a rating called METS (metabolic equivalent of task). The METS score is used in a formula that indicates the activity's power output in Calories/minute per kilogram of body mass. By conversion of body mass into weight in pounds and minutes into hours, we can interpret the Compendium's METS in units of Calories/hour.

$$\frac{\text{METS} \times \text{body weight in pounds}}{2.095} = \frac{\text{Calories}}{\text{h}}$$

Cycling at a speed of 20 mph, the Compendium assigns 15.8 METS.

The Compendium offers no guidance as to how to account for the weight of the bicycle; we will only use the weight of the rider (161 lb), since the machine's weight is not important at a constant speed on a level road. Doing the conversion from MET to Calories per hour, we obtain for the metabolic power:

$$\frac{15.8 \times 161\ \text{lb}}{2.095} = \frac{1214\ \text{Calories}}{\text{h}}$$

Table 8.3. Total muscular power to overcome air + rolling + bearing resistance.

constant = 0.006400 speed in ft/s, area =5.382 sq ft
rolling resistance of tires = 1.0 lb
bearing resistance = 0.00628 lb

Speed		Resistive force				Total mechanical power output				Total body power output—assuming a muscular efficiency of 20%			
mph	ft/s	air lb	rolling lb	bearing lb	total lb	ft-lb/s	watts	hp	Calories/hour	ft-lb/s	watts	HP	Calories/hour
2	2.93	0.06	1.0	0.00628	1.06	3	4.2	0.01	5	15.6	21.1	0.03	25
4	5.87	0.22	1.0	0.00628	1.23	7	9.8	0.01	11	36.0	48.8	0.07	57
6	8.80	0.50	1.0	0.00628	1.50	13	17.9	0.02	21	66.1	89.7	0.12	105
8	11.7	0.88	1.0	0.00628	1.89	22	30.0	0.04	35	111	150	0.20	175
10	14.7	1.38	1.0	0.00628	2.38	35	47.4	0.06	55	175	237	0.32	277
12	17.6	1.98	1.0	0.00628	2.99	53	71.4	0.10	83	263	357	0.48	416
14	20.5	2.70	1.0	0.00628	3.70	76	103	0.14	120	380	516	0.69	602
16	23.5	3.52	1.0	0.00628	4.53	106	144	0.19	168	532	721	0.97	842
18	26.4	4.46	1.0	0.00628	5.47	144	196	0.26	229	722	979	1.31	1143
20	29.3	5.51	1.0	0.00628	6.51	191	259	0.35	303	955	1296	1.74	1513
22	32.3	6.66	1.0	0.00628	7.67	247	336	0.45	392	1237	1679	2.25	1959
24	35.2	7.93	1.0	0.00628	8.94	315	427	0.57	498	1573	2134	2.86	2490
26	38.1	9.31	1.0	0.00628	10.3	393	534	0.72	623	1966	2668	3.58	3114
28	41.1	10.8	1.0	0.00628	11.8	485	657	0.88	767	2423	3287	4.41	3836
30	44.0	12.4	1.0	0.00628	13.4	589	800	1.07	933	2947	3999	5.36	4667
32	46.9	14.1	1.0	0.00628	15.1	709	962	1.29	1122	3544	4809	6.44	5612
34	49.9	15.9	1.0	0.00628	16.9	844	1145	1.53	1336	4219	5725	7.67	6681
36	52.8	17.8	1.0	0.00628	18.8	995	1350	1.81	1576	4976	6752	9.05	7879
38	55.7	19.9	1.0	0.00628	20.9	1164	1579	2.12	1843	5820	7897	10.6	9216
40	58.7	22.0	1.0	0.00628	23.0	1351	1834	2.46	2140	6757	9168	12.3	10699

Table 8.3 lists a power level at 20 mph of 1513 Calories/hour. The agreement between numbers calculated from basic physics (table 8.3) and real-world power measurements (the Compendium) is remarkable. Some discrepancy is to be expected due to variations in metabolic efficiency, cyclist frontal area, accuracy of measurements, etc. In particular, the assumed efficiency will have a major impact on the agreement of the numbers.

Another source of published data of cycling power requirements is provided in Professor Wilson's book *Bicycling Science* [5]. Wilson lists the power expenditures for a 187 lb rider/bike on a 'tourist bike' with a frontal area of 5.5 ft^2 riding at 16 mph as: tractive power = 149 watts and metabolic heat = 735 watts. If we consider the tractive power as equivalent to our mechanical power, the total power output and mechanical efficiency may be written:

$$(149\text{ watt} + 735\text{ watt}) = 884\text{ watt} = 760\,\frac{\text{Calories}}{\text{h}}$$

$$\text{mechanical efficiency} = \frac{149\text{ watt}}{884\text{ watt}} \times 100\% = 17\%$$

In spite of some variations in the size, frontal area, and muscular efficiency between Wilson's larger rider system (187 lb) and the 'standard rider' (181 lb system) in this book, the agreement with the power level of 842 Calories/hour at 16 mph in table 8.3 is reassuring.

8.11 Average speed versus average power

If I know my average speed, can I just use it to determine my average power?

Since an electronic cycle computer is an inexpensive accessory ($20–$30) that gives measurements such as distance traveled, ride time, average speed etc, many riders like to note and keep track of such parameters. It would appear a simple matter to use the average speed of a ride to infer the average power output. It is especially interesting to think of the power output in terms of Calories burned per hour. When viewing the power values in table 8.3, we must be careful to avoid a serious mistake. The last column, showing the power in Calories per hour, almost encourages this mistake. A glance at the table shows that riding at 12 mph requires a power expenditure of 416 Calories/hour, whereas riding at 24 mph demands a power output of 2490 Calories/hour.

Is it a simple matter to use the average ride speed from the cycle computer and read across table 8.3 to determine power expenditures, the Calories burned per hour? The short answer is no! The longer answer is a triple no!! One reason is that the cycle odometer knows nothing about the wind conditions. Are you riding into the wind or against the wind? The force of air resistance depends on the effective wind speed calculated as riding speed plus/minus wind speed. A similar argument can be made when riding up and down hills. The basic cycle computer knows nothing of slope forces.

What if you are riding with no wind on level ground—can you then use your average speed to get the average power? The answer is still no. If the ride involves any variation in speed, the average speed cannot be used to obtain an average power.

The power is the product of the exerted force times the speed:

$$P = F \times v$$

The primary resistive force is air resistance that depends on the square of the speed:

$$F \sim v^2$$

Thus, for the power required to fight air resistance:

$$P \sim v^3$$

The power is proportional to the cube of the speed. A speed increase from 12 mph to 24 mph requires eight-fold power increase to overcome air resistance!

Consider a numerical example for a two-hour bike ride. You start out easy, cycling at 12 mph for the first hour of the ride, and then pedal at 24 mph for the next hour. Your total distance is 36 miles over the two-hour period, an average speed of 18 mph. This is the reading of the bike's odometer and it is correct. Proper calibration of cycle odometers results in a distance accuracy of one per cent. Since the time measurement is also accurate, the average speed is very precise. The error arises when we use a simple average speed to compute the average power. Table 8.3 shows that at 18 mph the total body output is 1143 Calories/hour. Does this mean that for a two hour ride we burn 2286 Calories? The answer is no; however, it is one of those bad news, good news things. The bad news—you cannot use average speeds to get a true average power. The good news—your true average power will be higher than expected!

From table 8.3, the power for the first part of the trip (at 12 mph) is 416 Calories/hour and the power for the second half of the trip (at 24 mph) is 2490 Calories per hour. Thus, the total Calories burned is 2906 Calories—much better than the simple average result of 2286 Calories!

Only if your ride were at a constant speed would it be accurate to use the odometer's average speed to get the average power. If the ride consisted of segments, each one at a constant speed, it would be possible to do a power computation for each segment; you then total the power. The caution on wind speeds and hills still applies.

The most precise and complete way to determine power is to measure directly the rider force on the pedal. The pedaling force along with pedaling speed allows a computation of power ($P = F \times v$). Such technology does exist to measure the force and speed parameters and, thereby, determine instantaneous power expenditures. A summation of the instantaneous powers by the cycle computer allows determination of total power during the ride. Such technology is substantially more expensive than a basic cycle computer.

References

[1] Berto F 2009 *The Dancing Chain—the History and Development of the Derailleur Bicycle* 3rd edn (San Francisco, CA: Van der Plas Publishing) pp 375–7

[2] Henry F M; Demon J 1950 Metabolic efficiency of exercise in relation to work load at constant speed *J. Appl. Phys.* **2** 481–7

[3] Tokui M; Hirakoba K 2007 Effects of internal power on muscular efficiency during cycling exercise *Eur. J. Appl. Exercise* **101** 565–70

[4] Ainsworth B E *et al* 2011 The compendium of physical activities *Med. Sci. Sports Exercise* **43** 1575–81

[5] Wilson D G 2004 *Bicycling Science* 3rd edn (Cambridge, MA: MIT Press) p 76 table 2.1

Chapter 9

Temperature and heat

Some say the world will end in fire;
Some say in ice.

Fire and Ice by Robert Frost

Outdoor cycling is a quick lesson on the physics of temperature and heat. Temperature, along with other elements of weather such as moisture and wind, plays a major role in the cycling experience. To ignore these elements, the rider invites an encounter with discomfort and even disaster. The terms temperature and heat, common in daily conversation, need specific definitions and applications. This everyday use of the words is not misleading: temperature is related to a level of hotness or coldness; heat is something that can change the temperature. However, we need to refine these concepts further.

9.1 Temperature and its measurement

We begin with a precise, rigorous definition of temperature (that is not especially intuitive):

Temperature is a measure of the average kinetic energy (translational) of a body's molecules, atoms, and electrons

The molecules, atoms, and electrons are generically referred to as *particles*. By itself, this definition is of limited use. Let us think further—in chapter 8 we learned that kinetic energy is energy of motion; thus, temperature is a measure of the particles' motion—specifically, translational motion.

While this precise definition of temperature sounds very technical, it is consistent with our everyday experiences. Think of a winter morning when your hands feel cold—a common reaction is to rub the hands together to generate warmth. The reason the rubbing works is that the motion of our hands is transferred into motion of the molecules in the skin. Faster molecules have increased kinetic energy, and the temperature is a measure of this kinetic energy. For an even more dramatic example,

doi:10.1088/978-1-6817-4441-4ch9

bounce down a set of steps while sliding your hand along a wooden railing—your hand will get so hot it almost burns. As a third example—use your own judgment on this experiment—take a wire coat hanger and rapidly bend it back and forth a few times. Quickly touch the bent portion of the coat hanger to a sensitive region of skin. The coat hanger will be hot enough to make you say 'ouch'! In all of the above examples, the increased kinetic energies are manifested as a rise in temperature.

Our rigorous definition could be restated as 'temperature is an indication as to how fast the atoms, molecules, and electrons are jumping around'. Keep in mind that the jumping around refers to translational motion. Molecules can also exhibit other forms of motion such as spinning around and oscillating. The spinning and oscillating motions do not contribute to an object's temperature.

While there are various methods of measuring temperature, most rely on the fact that as an object's particles move faster they take up more room; hence, the object expands in size. Consider a classroom in which the students are sitting quietly at their desks, enthralled at the joy of physics. A room thirty-foot square might comfortably hold several dozen pupils. The bell rings; suddenly it is recess, and the students decide to play indoor rugby. As they run about crashing into each other and into the walls of the room, we might quickly conclude they need a larger playing volume. If the walls were elastic, the room would expand from the student collisions. Faster moving objects take up more room than slower moving objects.

Objects expand when their temperature is raised. The change in size allows for the development of a temperature scale and a thermometer. We might take a section of that metal coat hanger and use its length to determine the temperature. The absolute length of the wire is not important; cut it to any convenient size. The important factor is the change in the length of the wire in response to a change in temperature. Two temperature points are needed.

What temperature points should be the reference? It would be nice if these points are easy to reproduce—today, tomorrow or hundreds of years from now—and readily duplicated worldwide. For these reasons, water at two of its standard phase transitions is used. The phase transitions occur at the temperature when water turns from liquid to solid ice and the temperature when water turns from liquid to gas (steam).

The basic experiment is simple—use two glass containers—one with a mixture of ice and water (the exact percentage is not important), the other with water heated until boiling with white stuff rising upward (figure 9.1). Notice, we did not say steam is appearing—steam is the gaseous (i.e. vapor) state of water—individual water molecules. You cannot see these individual molecules. The white stuff rising from a boiling pot is water droplets—molecules that have already condensed back into the liquid state.

The technique is straightforward—plunge the metal bar into the ice water; give a few moments for it to stabilize in temperature and size. Once the bar has stabilized, measure its length. Give this length a number—an arbitrary number. Next, repeat the experiment at the boiling point with the same metal bar. Assign the boiling point another arbitrary number. In essence, a similar procedure was performed by Daniel Gabriel Fahrenheit which resulted in 32° for the ice water mixture and 212° for the boiling point. Another temperature scale, named after Anders Celsius, uses numbers 0° for ice water and 100° for boiling water.

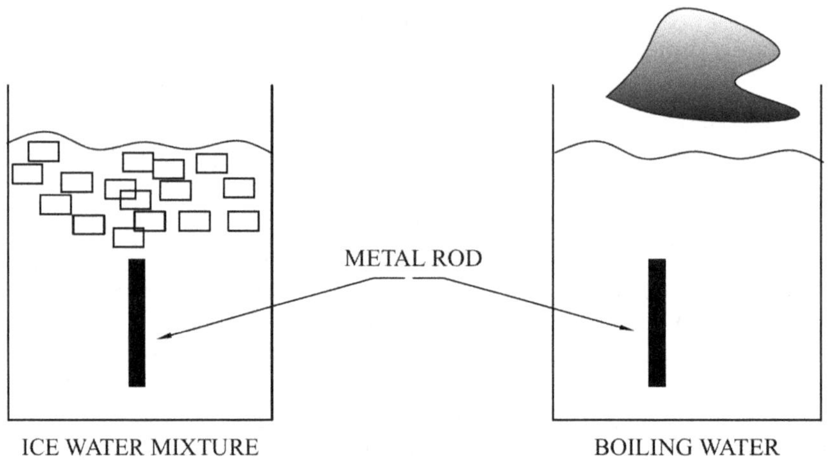

Figure 9.1. Temperature measurement.

Are all temperature scales so arbitrary? Yes, except for one—the Kelvin scale. The zero (called absolute zero) in the Kelvin scale is approximately −273 °C (−459 °F). While many cyclists enjoy cold weather riding, few will be riding at absolute zero!

9.2 Heat

> Keeping in mind that temperature is a measure of atomic, molecular, and electronic motion, we regard heat as the energy transferred to or from a body that changes the body's atomic, molecular, and electronic motion. Physics teachers get a bit fussy here—strictly speaking, the term 'heat' should be used only when the energy is in transit. In other words, a pizza right out of the oven is hot, not because it contains a lot of heat; rather it is hot because there is a high level of atomic, molecular, and electronic motion. If you take a bite of the hot pizza and it burns your mouth, it is because heat was transferred from the pizza to your mouth. The delivered heat then raises the temperature of your mouth and you wince.

Consider another example—take two bricks; the first is placed in an oven and heated to 400 °F and the other is placed in a freezer and cooled to 0 °F. The two bricks are then put into an idealized insulated chamber; assume the chamber has no heat transfer with the bricks (figure 9.2A).

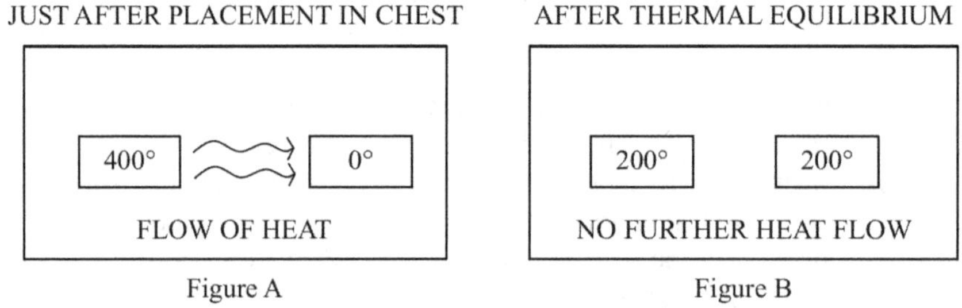

Figure 9.2. (A,B) Bricks in insulated chest.

The hot brick contains an intense level of particle motion and the cold brick has little motion. After being placed in the chest, the hot brick will cool down and the cold brick will heat up. These changes in temperature are due to *heat flowing* from the hot brick to the cold brick. Let us be precise; the heat only exists while it is in transit—flowing from one body to another. Eventually, if identical bricks are left to stabilize in a perfectly insulated chamber, where there are no heat losses or gains with walls of cooler or the outside surroundings, they will come to thermal equilibrium at 200 °F (figure 9.2B).

The flow of heat normally occurs from warmer objects to cooler objects. The flow is analogous to a ball rolling on a hill. Usually, the ball rolls from the top of a hill to the bottom of a hill. Can a ball travel from the bottom to the top? Of course, when work is done on the ball. Similarly, the performance of work causes heat to flow from a colder object to a warmer object. This is the situation in your refrigerator or freezer. The flow of heat in the reverse direction is due to the work being done by the electric motor driving the compressor. If you put your hand behind or under the refrigerator, you will notice the warm air associated with the exhaust of heat from the cooler interior to the warmer surroundings of the kitchen.

9.3 Units of heat

There was a time when heat was considered a unique fluid; it was called the 'caloric'. The caloric flowed from a hot body to a cold body. In the latter half of the nineteenth century, it was recognized that heat is best viewed as a form of energy. Nevertheless, we often refer to the 'calorie' as a unit of measurement for heat.

A calorie of heat is defined as the amount of heat required to change the temperature of one gram of water by one degree Celsius.

A gram of water has a volume of one cubic centimeter, about the size of a fingertip. A related unit is the Calorie equal to 1000 calories. In everyday usage, such as the calorie content of food, the capitalized Calorie is intended; e.g., 'the serving of cake contains 450 Calories'. Note: the reader will observe in this chapter the use of metric units. Even in the United States with its entrenched Customary units, the Calorie is the most familiar unit of energy.

9.4 Heat generation on a bicycle

It is easy to develop an estimate for the body's cooling requirements at a particular level of cycling. Think about our standard rider pedaling to overcome the forces of air, bearing, and rolling resistance (chapter 5). In section 8.10, we learned that a rider operating at a 20% muscular efficiency must exert a total body power output equal to five times the required mechanical power; table 8.3 shows that at a speed of 20 mph (no wind, level road) the total body power output is:

$$P = 1296 \text{ watts} = 1513 \frac{\text{Calories}}{\text{hour}}$$

Twenty percent of this power goes into moving the rider and bicycle; the other 80% of the rider's power:

$$0.80 \times 1513 \frac{\text{Calories}}{\text{hour}} = \frac{1210 \text{ Calories}}{\text{hour}} \tag{9.1}$$

results in internal heat generation within the cyclist.

A very significant amount of body heat generated during the ride! In order that the cyclist not overheat, this heat energy must be removed from the body through a cooling process.

9.5 Mechanisms for heat transfer

How do we keep our bodies from overheating while at rest and while riding a bicycle?

In this section, we examine the mechanisms for the transfer of heat between objects. These mechanisms for heat transfer are very important to a cyclist riding in warm weather. The concern is to avoid overheating from the large amount of waste heat generated. Conversely, when riding in cold weather, the cyclist must guard against excessive heat loss caused by the moving air.

Traditionally, most physics books list four mechanisms by which heat is transferred. They are: conduction, convection, radiation, and evaporation.

The basic principle of each mechanism is examined and applied to the cyclist. Most mechanisms for heat transfer depend upon the temperature difference between objects. In considering the heat transfer from a cyclist, one of the objects is the surface of the rider's body and the second object is the environment. While it is often stated that the 'normal' human body temperature is 98.6 °F, this temperature refers to the internal core temperature; the exterior surface temperature will be much lower. In our analysis, we will use a cyclist skin surface temperature of 91 °F(33 °C).

9.6 Conduction

Conduction is caused by collisions between atoms, molecules, and electrons

Suppose one side of an object is at an elevated temperature caused by a high level of particle motion; these fast moving particles will collide with their slower neighbors and transfer the motion. It is helpful to visualize the transfer of energy as a row of dominoes falling down. The motion of a falling domino is transferred to its neighbor which then falls and transfers the energy to the domino next in line.

Imagine sitting around a campfire. Perhaps, you have an iron poker; instead of just moving the logs around, take the poker and bury one end into the red-hot embers. Continue to hold the other end with a bare hand. After holding the poker for a few minutes, notice that the upper end gets too hot to handle. Heat has been transferred from the red-hot coals to your hand by means of conduction. The particles in the red-hot embers crash into the particles at the bottom of the poker; these collisions continue up the length of the poker until the fast moving particles are in contact with the skin. Eventually, the molecules in the skin pick up the motion and begin to move fast. Your hand is hot!

For effective conduction, there must be many closely spaced particles that are free to move about and near enough to crash into one another. Metals are good heat conductors due to their electrons' ability to move freely. In contrast, a material like air is a very poor conductor. Although the air molecules are free to move, they are not in close contact. We readily push the air aside as we move about!

An empirical expression for heat conduction is written as follows:

$$\text{rate of heat transfer} = \frac{Q}{\Delta t} = K \times \frac{A}{L} \times \Delta T \tag{9.2}$$

where:

- $Q/\Delta t$—is the rate of heat flow, often expressed in units such as Calories/hour.
- A—is the area of the surface through which the heat flows (m^2).
- L—is the thickness of the conducting material (m).
- ΔT—is the difference in temperature between the warm and cold objects (degrees Celsius).
- K—is the thermal conductivity of the substance; we list a few sample values in the table 9.1 [1][1].

Table 9.1. Values of thermal conductivity.

Material	k (in Calories/h-m-°C)
copper	342
air	0.0206
water	0.530
iron	69
aluminum	204
carbon fiber[a]	128

[a] Packard A, Reynolds Cycling, personal communication

As expected, the metals, with their free moving electrons, have high thermal conductivity. The relatively low thermal conductivity of carbon fiber might be surprising. Given the fact that carbon is a very good conductor of electricity, why does carbon fiber not have a high thermal conductivity? The carbon fiber is a composite of strands of carbon held together with an epoxy resin. The epoxy binder is a poor conductor and, hence, it lowers the overall thermal conductivity of the carbon fiber matrix. This poor thermal conductivity of carbon fiber results in an issue with carbon wheels during long descents.

How effective is conduction in the removal of heat from a cyclist? The above equation (9.2) is more difficult to apply to the human body than to the hot poker or insulation in the walls of a house. The poker has an obvious length and the walls have a known thickness. A moving body is an additional complication. Nevertheless, we try for an estimate:

Surface area: articles in the literature suggest formulae for total body area based on height and weight [2]. For a 5'10" rider of weight 161 lb, the surface area works out to about 1.90 m^2.

[1] www.The Engineering Toolbox/thermal conductivity

Temperature difference: on a warm day, this temperature difference will be small. The skin temperature of a human is about 91 °F (33 °C); for air temperature, we use 81 °F (27 °C). Thus: $\Delta T = 6$ °C. This is a very small difference in temperature. Contrast this difference to that of a poker tip embedded in red-hot coals at ~1500 °F.

Thickness of the conducting material: we are referring to the thickness of the layer of air between skin at 91 °F and the ambient air temperature at 81 °F; this factor is difficult to estimate. In order to get a feeling for the magnitude of the conductive effect, estimate a thickness of one inch (0.0254 m). Another consideration in looking at conduction is the effect of clothing. Assume that with warm weather cycling skins, the thin layer of fabric does not appreciably affect the conduction of heat from the cyclist's body.

Substituting the numbers into (9.2):

$$\begin{aligned}\frac{Q}{\Delta t} &= K \times \frac{A}{L} \times \Delta T \\ &= 0.0206 \frac{\text{Calories}}{\text{h–m–°C}} \times \frac{1.90\ \text{m}^2}{0.0254\ \text{m}} \times 6\ \text{°C} \\ &= \frac{9.25\ \text{Calories}}{\text{hour}} = 10.8\ \text{watts}\end{aligned}$$

The conductive heat loss is not significant. The body's large surface area for heat flow is offset by the air's low value of thermal conductivity and the small temperature difference between the rider's skin and the ambient air. Riding in cold conditions of 32 °F (0 °C) will result in a much larger temperature difference and a heat loss of approximately five times the warm weather loss.

9.7 Convection

Convection is a form of heat transfer that occurs when warm fluids rise

To understand convection recall that, as the temperature of an object is raised, the particles move faster and the size of the object increases. Consider a parcel (an imaginary balloon) of air sitting just above a hot stove. This parcel of air is heated via conduction (figure 9.3A). As the temperature of the parcel increases, the parcel expands and becomes less dense (chapter 3). This heated air, with a lower density than the surrounding air, will now rise upward due to a buoyant force (figure 9.3B). The effect is similar to the buoyant force on a helium balloon.

Air is a very good convector; it easily heats up, expands and the parcel of air freely rises. Have you ever noticed warm air near the ceiling of a heated room? Consider a room with a source of heat on one wall—a baseboard heater or a potbelly stove—the warmed air above the heater rises and travels along the ceiling (figure 9.3C). As it cools, it usually drops on the far side of the room. The cool air is then drawn across the floor back toward the heater. The draft of cool air can be quite noticeable; you feel chilled. We will shortly discuss cooling by evaporation and see that moving air increases evaporative cooling. The circulating parcels of air are called 'convection cells'.

Convection increases the effectiveness of conduction. In the above discussion, the air close to the top of the stove was heated by conduction. If the warmed parcel were

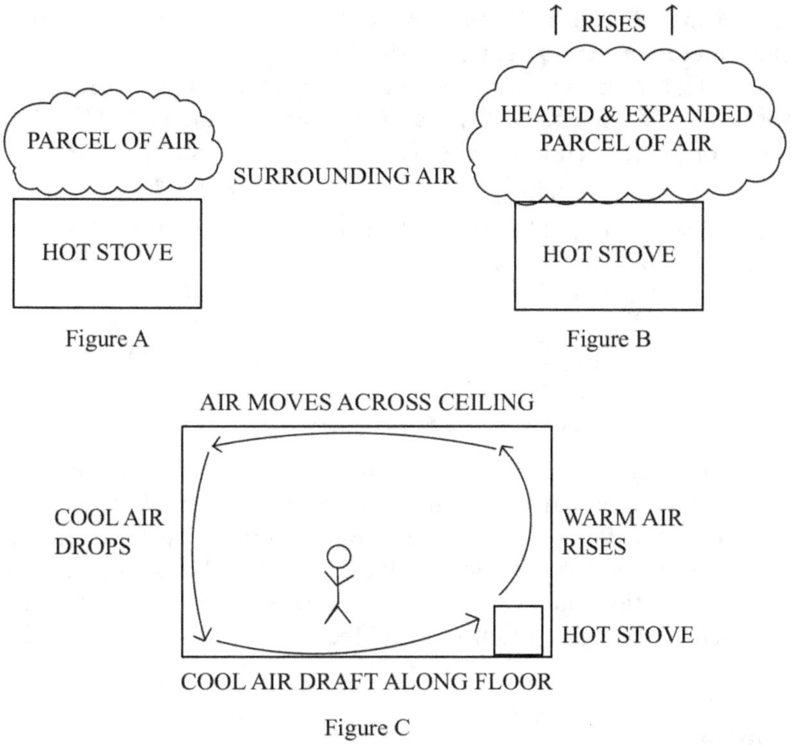

Figure 9.3. (A,B,C) Convection.

constrained in one place, it would eventually reach the temperature of the stovetop. Think of convection as a way for warmed air to 'escape' and move to the ceiling. Convection serves to 'carry away' the warmed air. In the room's convective circulation, new cooler air is moved into place above the stove, thereby, maximizing the conduction of heat from the stove. Convection has such a major impact on the conductive method of heat transfer that the two mechanisms are often considered together under the heading *convection*. An empirical model for convection of heat from the human body has been proposed by Wegner [3]. In this model, the rate of heat removal may be written:

$$\frac{Q}{\Delta t} = h_c \times A \times (T_s - T_a) \tag{9.3}$$

where:

$Q/\Delta t$—is the rate of heat flow, often expressed in units such as Calories/hour.
A—is the area of the surface through which the heat flows (m^2).
h_c—is a proportionality constant that varies with the air speed past the body. The air speed is the sum of the rider's cycling speed and the atmospheric wind.
T_s and T_a—are respectively the temperatures of the body surface and the ambient air.

An important aspect in the above equation is that convective heat flow is proportional to the difference between the rider's skin and the air temperature. If the day is warm with the air temperature at or above the rider's body surface temperature, there will be no heat removal due to convection.

To get a feeling for the magnitude of convection, let us calculate the convective heat transfer for a rider of skin area 1.90 m^2 and skin temperature of 33 °C (91 °F) cycling on a nice day of air temperature 23 °C (73 °F). The values of h_c, extracted from the reference article, are for air speeds that are on the low side for a cyclist; the highest air speed in the article is 5 m/s (11.2 mph). At a speed of 5 m/s, $h_c \approx 31 \frac{\text{watts}}{\text{m}^2\ {}^\circ\text{C}}$ thus:

$$\frac{Q}{\Delta t} = h_c\, A(T_s - T_a)$$

$$= 31 \frac{\text{watts}}{\text{m}^2\ {}^\circ\text{C}} \times 1.90\ \text{m}^2 \times \left(33\ {}^\circ\text{C} - 23\ {}^\circ\text{C}\right)$$

$$= 589\ \text{watts} = 506\ \text{Calories/h}$$

We see that, on a pleasant cool day, convection can remove a significant amount of the cyclist's heat. A problem arises when the day is warmer. For instance, at an air temperature of 32 °C (90 °F), the convective heat removal will be only ten percent (59 watts, 51 Calories/h) of the above values. Riding at higher speeds will increase the effect of convection but we emphasize that, when air temperature equals or exceeds skin temperature, convection is ineffective as a body cooling mechanism.

9.8 Radiation

While the word 'radiation' sounds bad and the phrase 'waves of electromagnetic radiation' sounds even worse, we will see that there are both nice and not so nice types of radiation. First, consider something very nice—a quiet, still lake on a warm summer day. Toss a small rock into the lake and observe the ripples as they travel to shore. The ripples are a commonplace example of waves. Should the waves pass a fishing bobbin, you will notice the bobbin moves vertically up and down. The motion of the water is vertical but this motion, the wave, travels horizontally to shore. If you look carefully, you will note that the wave consists of hills (crests) and valleys (troughs). The distance from one peak to the next is the *wavelength*—maybe a few inches for this water wave.

In the physical world, there are many other types of waves—sound, a rope shaken at one end, and an important category called *electromagnetic (em) waves*. What are these em waves? Play with two magnets—notice how they repel and attract each other. These forces of repulsion and attraction occur via a *magnetic force field*. In other words, the first magnet emits a force field that travels to the second magnet and produces the attractive or repulsive force. A second example of a force field is the *electric force field* that occurs between two electric charges. These fields might seem very mysterious and difficult to visualize, but there is another field with which you are very familiar—the Earth's gravitational field. Fields are not so mysterious after all.

Radiation is a form of heat transfer that occurs with electromagnetic (em) waves

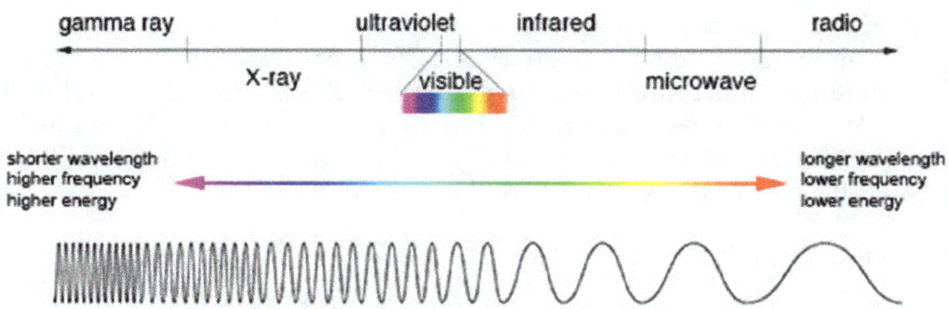

Figure 9.4. Electromagnetic spectrum. Courtesy NASA image the universe.

What is an electromagnetic wave? It is a pair of fields, one electric and the other magnetic, that vibrate and travel like a wave. These em waves are common and easy to produce; turn on a flashlight and shine it across the room. If the light hits a wall, we see the shape of the beam. Exactly what comes out of the flashlight? It is a wave of electric and magnetic fields! A display showing electromagnetic waves arranged according to their wavelength is called the *electromagnetic spectrum* (figure 9.4).

These electromagnetic waves encompass radiations that range from radio waves on the long wavelength side down to x-rays and gamma rays on the short wavelength side. Some of these electromagnetic waves trigger sensations in the eye's retina and are called *visible light*. Even the colors of visible light differ in their wavelengths; the reds are the longest visible wavelengths and the violets are the shortest visible wavelengths.

The waves just longer than red light are called *infrared*. Our skin perceives the infrared radiations as *heat*. Think about sitting around a campfire, a fire that has been burning for some time with a nice glowing bed of red-hot coals. The heat you feel on your face is the infrared electromagnetic waves traveling from the coals to your face. The coals are *emitting* electromagnetic radiation.

For a body emitting radiations, the amount and specific wavelengths depend upon the body's temperature. As with conduction and convection, the radiative energy transfer from a body depends on the temperature of the body and the temperature of the surroundings. We get an estimate for the heat transfer via radiation by means of a relationship called the Stefan–Boltzmann equation.

$$\frac{Q}{\Delta t} = e \times \sigma \times A \times (T_s^4 - T_r^4) \tag{9.4}$$

e—is called the emissivity; it is a measure of how well objects absorb and emit radiation. For skin, the emissivity is close to 1.0.

σ—is called the Stefan Boltzmann proportionality constant; it is equal to:

$$\sigma = 5.67 \times 10^{-8} \frac{\text{watts}}{\text{m}^2 - \text{K}^4}$$

T_s and T_r—are the temperatures of the rider's body, and the surrounding temperatures. These temperatures must be expressed in the Kelvin temperature scale; the conversion is simple: degrees K = degrees C + 273. For the rider's skin temperature (91 °F, 33 °C): T_s = 33 °C = 306 K and for the surrounding air temperature: (81 °F, 27 °C): T_r = 27 °C = 300 K.

A—is the area of the rider's skin surface; again we use 1.90 m^2.

We substitute into (9.4):

$$\frac{Q}{\Delta t} = e\,\sigma\, A(T_s^4 - T_r^4)$$

$$= 1.0 \times \left(5.67 \times 10^{-8} \frac{\text{watts}}{\text{m}^2 - \text{K}^4}\right) \times 1.90\ \text{m}^2 \times (306\ \text{K}^4 - 300\ \text{K}^4)$$

$$= 71.9\ \text{wattts} = 61.9\frac{\text{Calorie}}{\text{h}}$$

While this magnitude of the radiative heat loss is significant, there is an important additional consideration for an individual exercising outdoors. In the calculation above, we used surroundings (27 °C) that are cooler than the skin temperature (33 °C). The human body was radiating energy to its surrounding environment. What happens when the surroundings are warmer than the rider's skin temperature? In this case, the radiative process causes the heat to flow into the rider! Even in situations of cool air temperature, surfaces such as dark road pavements can be quite warm from sunlight. Riding in direct sunlight exposes the cyclist to vast amounts of radiative energy from the Sun. Overall, for a cyclist riding outdoors on a warm day, the radiative method of heat transfer is more often a heating rather than a cooling process.

9.9 Evaporation

Why is it necessary to drink so much water on a long bike ride?

The importance of drinking adequate amounts of water and maintaining proper hydration during strenuous exercise is well known. Along with serious falls and crashes, dehydration and consequential heat exhaustion and heat stroke are at the top of cycling hazards. In this section, we explore the physics of water and the role it plays in keeping the body from overheating. Begin by thinking about a simple experiment. Consider a pot containing a mixture of water and ice that is placed over a steady source of heat; a stove set to high heat is ideal. The exact percentage of ice and water is not important; a fifty-fifty mixture is good. Also needed is a thermometer to measure the temperature of the mixture and a clock to measure time; it is best to keep the mixture stirred. The experiment is very easy; every minute or so, read the thermometer and observe what is happening in the pot. There is not a direct way of measuring how much heat is added to the pot but, since we are using a steady source of heat, we can consider the elapsed time to be a measure of the heat added to the pot. After collecting the data of temperature versus time, a graph will help visualize the results (figure 9.5 A).

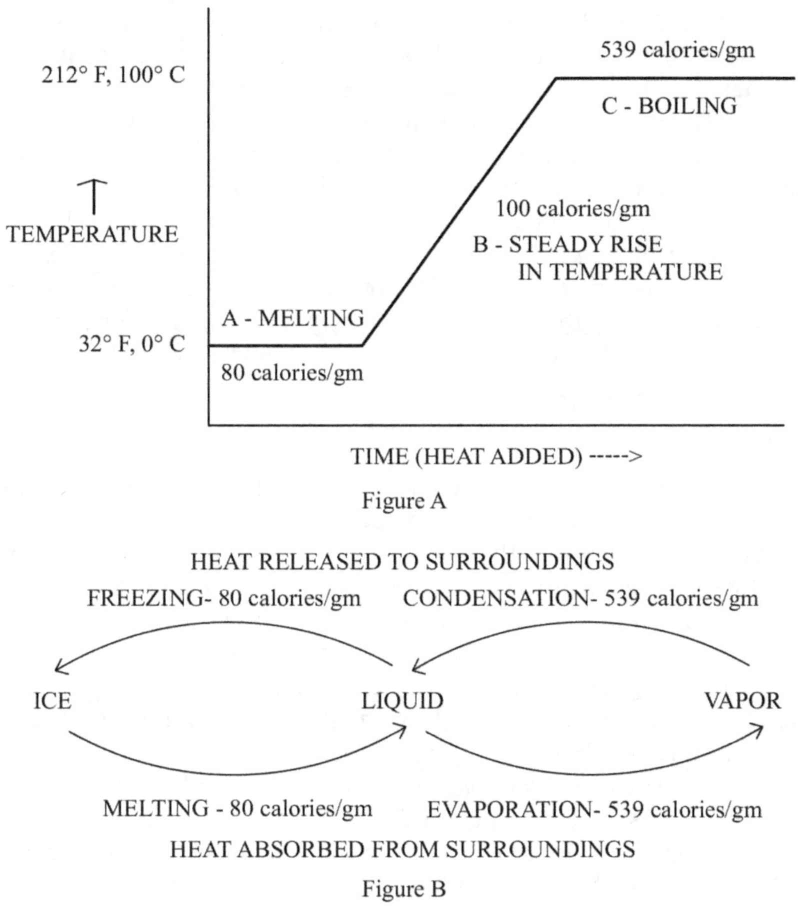

Figure 9.5. (A,B) Phases of water.

Since we begin with an ice-water mixture, the starting temperature should be 32 °F (0 °C); recall the discussion of temperature scales in section 9.1. What will be the temperature after one minute? Is there any ice left? If so, the temperature will still be 32 °F. The temperature will stay fixed at 32 °F for as long as there is ice and the pot is properly stirred. The length of the period of fixed temperature depends upon the amount of ice, water, and the heat delivered by the stove. Observation of the contents of the pot shows the ice slowly melting. Call this period, when the temperature is fixed at 32 °F and the ice is slowly melting, region A in the graph.

Eventually, the last bit of ice melts and there is now a steady increase in temperature. Call this period of rising temperatures region B.

When the temperature climbs to 212 °F, we observe white stuff coming from the pot (we are trying to avoid calling this white stuff 'steam'). In addition, bubbles are forming and rising to the top of the pot—the water is boiling. Suppose the pot sits there boiling for a while; the temperature is seen as fixed at 212 °F. This portion of the experiment will be referred to as region C.

In the three regions of the graph, the water is transitioning between its three phases—solid (ice), liquid (water), and gas (steam i.e vapor). The heat added to the pot is the energy source driving these phase transitions.

Region A—The ice is *melting*. In the frozen state, the water molecules are packed close together in a highly ordered crystalline arrangement. The water molecules in this low energy state are happy to sit in one position next to each other. As the heat is transferred into the ice, the energy serves to break up this close packed arrangement. In an introductory physics experiment, the amount of heat (called heat of fusion) required to melt ice is measured to be 80 calories per gram of ice. Eighty calories of heat must be absorbed by the ice to convert one gram of ice into one gram of water. After absorbing these 80 calories and melting, the water will still be 'ice cold' at the freezing point of 32 °F (0 °C). The heat of fusion is a two-way street; when water at 32 °F (0 °C) freezes into ice at 32 °F (0 °C), 80 calories per gram of heat are released to the surroundings (figure 9.5B).

Region B—The temperature of the liquid is steadily *increasing* toward the boiling point. The heat transferred into the water is serving to increase the average translational kinetic energy of the molecules. If you could observe the molecules, you would see they are moving faster and faster as the temperature climbs. However, because the water is in the liquid phase, the molecules are still close and touching one another as they move about. Recall that the definition of the calorie is the amount of heat required to change the temperature of one gram of water by one degree Celsius. Therefore, to heat water from its freezing point to its boiling point requires 100 calories per gram (figure 9.5A).

Region C—The water is *evaporating*. The pot of water is observed to be boiling and steam is generated. As heat is continuously added to the pot, the energy is acting to break the individual water molecules free from the bonds that hold one molecule to another. As the separated molecules break free, they jump out of the pot into the air. Since water molecules are strongly attracted to each other, a sizeable amount of energy is needed to separate them. The heat required to do so is called the *heat of vaporization*. An elementary physics lab experiment can determine the heat of vaporization; at the boiling point, it is found to be 539 calories per gram of water. This heat of vaporization is also a two way process—heat is absorbed by the water as it is turned into a vapor and heat is released by the vapor as it condenses back into liquid (figure 9.5B).

The evaporation of water removes a substantial amount of heat from the system!

Of course, an open container of water will evaporate without being raised to its boiling point. Envision the water molecules at the surface. Since the molecules are moving about with a range of speeds (recall temperature is a measure of the average motion), there will always be some molecules that exceed the average. The fastest moving molecules near the surface will escape from the bucket. Little by little, perhaps over many days, the water molecules jump out of the bucket; eventually we find the bucket empty and say the water has *evaporated*. This slower process of evaporation at less than the boiling point also involves a heat of vaporization. At a skin temperature of 91 °F (33 °C), a gram of water requires 577 calories to evaporate. The extra energy beyond the 539 calories/gram is a consequence of the water molecules not moving as fast at the lower temperatures as they do at the boiling point; thus, more energy is required to free them from the liquid phase.

9.10 Cooling effects of evaporation

How does sweating keep us cool?

Strenuous work and exercise produces sweat on the surface of the skin. The presence of water on the skin serves to remove heat from the body through the evaporation mechanism. Evaporation is the primary mechanism for removing heat from the body—as we exercise, we work up a sweat, and our skin gets wet! At a skin temperature of 33 °C, the evaporation of the water from the skin absorbs 577 calories per gram of evaporated sweat from the surroundings—the surroundings primarily being the exercising body.

In addition to the evaporation of moisture from wet skin, another mode for evaporative dispersal of body heat is through panting. The vigorous exhale moves a substantial amount of vapor from the lungs to the atmosphere. This vapor carries heat from the body.

> Section 9.4 showed that our standard cyclist, riding at 20 mph, generates 1210 Calories/hour of excess heat. Since conduction, convection, and radiation do not provide sufficient cooling at skin temperatures, the primary means of cooling during this strenuous exercise is evaporation. When air temperature is above skin temperature, conduction, convection, and radiation perform no cooling function; rather, the warm air serves to add heat to the exercising body. In this situation, the only mechanism for the body to eliminate heat is evaporation.

How much water should I drink to avoid overheating during a long ride?

Although the body can withstand a short period of dehydration, it is interesting and straightforward to estimate the amount of water needed to evaporate the excess heat of 1210 Calories/h from the cyclist. We use a heat of vaporization of 577 calories per grams of water (do not forget to convert Calories to calories):

$$1210\frac{\text{Calories}}{\text{hour}} = 1\,210\,000\frac{\text{calories}}{\text{hour}} \times \frac{1 \text{ gram of water}}{577 \text{ calories/gram}} = 2097 \text{ grams of water}$$

This amount of water converts to 2.1 liters or 2.2 quarts. Approximately, three large water bottles per hour!

Is it appropriate to say that dripping sweat is wasted sweat?

To achieve the cooling effect, the water must evaporate. If the sweat drips or is blotted off the body, it does not get a chance to evaporate and remove heat from the body. Do not waste your sweat!

Besides sweating, are there any other ways that water keeps the rider from overheating?

The benefit of pouring water on the surface of the body is obvious. We can think of the water as 'free sweat' without the danger of dehydration. We can also envision panting (heavy breathing) as a mechanism for the body to expel water vapor. This vapor is carrying heat from the body's core.

Why do I sweat more when I stop riding?

You probably do not sweat more but, during a riding break, you lose the breeze that aided in the evaporation of the water. Therefore, when you stop, you will notice the sweat build up on your skin. The overheated feeling you experience during the break quickly turns into a nice cooling effect once you resume pedaling.

9.11 Role of cycling clothing

Is there a valid purpose for those skintight clothes, or are they just supposed to make you look cool?

Whether a particular rider looks 'cool' in spandex is beyond the scope of this book. Looking cool is one thing; staying cool—or at least not overheating—is another. An implication of the evaporation mechanism is that the water vapor must be allowed to travel away from the body. Recall that the opposite of evaporation is condensation and that the condensation of water vapor releases heat to the surroundings. If the vapor condenses back into liquid water near the body, it tends to negate the cooling process. We now appreciate the importance of wearing clothing that allows the water vapor to pass to the atmosphere away from the skin. If you find that your clothing is excessively wet after a hard ride, the water that evaporated and removed heat from your skin then had a chance to condense and release heat in the vicinity of your body. The cooling effect of the evaporation has been compromised. Given the importance of evaporative cooling, it is apparent that riding in waterproof clothes is very bad idea. The ultimate in exercise clothing is made from modern fabrics that utilize micro pores. These tiny pores allow the passage of the small water vapor molecules from the body while blocking the infiltration of the larger drops of liquid water such as rain.

9.12 Wind effects on cooling capacity of evaporation

He sends cold northern blasts
that harden the ponds like solid ground,
Spreads a crust over every body of water,
and clothes each pool with a coat of armor.

New American Bible, Book of Sirach, chapter 43, Verse 20

Whether a body is sweating profusely or not, the evaporation of water from the skin has a major effect in removing heat from the body. It is the primary reason for the *wind chill effect*. Every cyclist has experienced feeling much colder while riding a bicycle than when at rest. Moving air past the rider assists in the removal of heat

Table 9.2. Water vapor capacity for a parcel of air at standard pressure.

Temp. (°C)	Temp. (°F)	Water vapor capacity (Grams of water vapor per kg of air (g/kg))
0	32	3.5
5	41	5.0
10	50	7.0
15	59	10.0
20	68	14.0
25	77	20.0
30	86	26.5
35	95	35.0
40	104	47.0

from the body. In addition to enhancing evaporation, the movement of air increases other cooling processes such as conduction and convection. The cooling effect and specific mechanisms are an active area of research [4, 5]. There are various versions of 'wind chill' charts that allow for an estimate of perceived temperature versus the actual temperature and wind speed. Regardless of the theories and scientific charts, riders quickly learn the level of clothing that affords the desired comfort level.

Can we estimate the cooling power of this wind driven evaporation? The evaporation of water from a surface is not trivial to model; factors such as wind, atmospheric pressure, temperature, and the amount of existing water vapor in the atmosphere (humidity) all play a role. Interesting models have been developed to calculate evaporation losses and heating requirements of swimming pools, fish farms, etc. The resulting empirical equations offer some estimates on evaporative heat losses from wet surfaces[2]. The swimming pool model indicates that a cyclist, traveling 20 mph with fully wetted skin, can experience a wind chill effect of near 3000 Calories per hour (author's unpublished result). At high levels of humidity, the evaporative cooling power is reduced by a factor of tenfold; the body's major cooling mechanism is seriously compromised.

9.13 Humidity and dew point

Is it really 'not' the heat but the humidity?

The efficiency of the body's evaporative cooling process is strongly dependent on the amount of water vapor that exists in the atmosphere. In this section, we consider the ability of a parcel of air to hold water vapor and the dependence on air temperature. While the molecules of atmospheric water vapor would like to stick to each other and form droplets, there is a counter tendency for fast moving molecules to bounce off rather than stick together. Table 9.2 shows, at a range of temperatures, the amount of water vapor (in grams) that a given parcel of air (in

[2] www.engineeringtoolbox.com/evaporation-water-surface-d_690.html

kilograms) can hold. Although this table refers to water in its vapor state, a gram of water is about the size of fingertip; a kilogram (2.2 lb) of air is found inside a very large beach ball (chapter 3).

From table 9.2, we note that the capacity of a parcel has a strong dependence on temperature. Air at 59 °F can hold 10 grams of water vapor per kilogram of air; at 41 °F the parcel can only hold 5 grams of water vapor. This temperature dependence is a consequence of the water molecules moving faster at higher temperatures and thus more likely to bounce off rather than stick together and form droplets. On a muggy day, you will readily observe water droplets forming on the outside of a drinking glass containing a cold liquid.

The capacity numbers in table 9.2 indicate the maximum amount of water vapor that a parcel can hold without condensing into droplets. Most of the time, the actual amount of water vapor in the parcel will be less than the capacity values of table 9.2. Consider an example of a parcel at 77 °F that has an *actual water vapor* of 15 gm/kg. Since the capacity at 77 °F is 20 gm/kg, there is 'room' in the parcel for additional water vapor. The ratio of actual water vapor to the parcel's capacity is the basis for the definition of *relative humidity*:

$$\text{Relative humidity} = \frac{\text{actual water vapor}}{\text{capacity}} \times 100\%$$

With the 77 °F parcel containing an actual water vapor of 15 gm/kg, we obtain for the relative humidity:

$$\text{Relative humidity} = \frac{15}{20} \times 100\% = 75\%$$

The temperature and relative humidity have an inverse relationship. Suppose in the example above, the temperature rises with no change in the actual water vapor. This might occur as the air is simply warmed during the afternoon. At an afternoon high of 86 °F, the capacity is 26.5 gm/kg and the relative humidity is only 57%. As the Sun sets, the temperature might drop to 68 °F. The parcel at 68 °F will have a capacity of only 15 gm/kg and the relative humidity is now 100%.

Since evaporative cooling is very important in removing heat from an exercising body, riding a bike in the afternoon environment of 86 °F, 54% humidity might be more pleasant than the clammy conditions of 68 °F, 100% humidity.

The oft repeated exercise mantra of 'listen to your body' is especially appropriate when cycling in warm weather. The rider is well advised to not only pay attention to how she feels but also how the brain is reacting. Often, heat stress first affects the mind. Cycling at high speeds requires constant vigilance for road conditions and hazards such as pedestrians and vehicles. If you find yourself making small judgmental mistakes, such as dropping a tire off the road edge or hitting small stones and branches, it may be a sign that your mind has lost a critical level of sharpness.

9.14 Specific heat

Have you ever been hungry—really hungry—and taken a big bite of a hot slice of pizza, fresh from the oven, and managed to burn the roof of your mouth? Sometimes, you do not learn and proceed to take a second bite, burning your mouth again. What to do? Take a gulp of cold soda, blow on the pizza, wave it around? None of these options will satisfy your ravenous craving for eating that tasty, mouth-watering pizza and you might even be asked to leave that fancy pizza parlor. Next time, with your twice-burned mouth, take a bite from the crust. It will not burn your mouth. What is the difference between the front of the pizza with juicy cheese, tomato sauce, pepperoni, etc and the back portion of dry crust? Both portions of the pizza were probably at the same 400 °F temperature when they came out of the oven. The difference is a physical parameter called the *specific heat*, sometimes called the *heat capacity*.

The amount of heat Q to change the temperature of the object is proportional to several factors: m is the mass of the object, ΔT is the change in temperature, and c is called the specific heat

We write an expression:

$$Q = c \times m \times \Delta T \tag{9.5}$$

In the CGS metric system, where heat is measured in calories, temperature in Celsius, and mass in grams, the units of specific heat are calories per gram—degrees C.

It is best to think of *specific heat* as the number of calories of heat required to change one gram of a substance by one degree Celsius. Recall in section 9.3, the calorie was defined as the quantity of heat required to change the temperature of one gram of water by one degree Celsius. This implies that the specific heat of water is 1.0. How does this compare to other substances? Most common materials have a specific heat significantly less than that of water; a few are listed in table 9.3 [6].

Water's high specific heat is due to the water molecules being strongly attracted to each other; the tendency of water to 'bead up' and form drops is a consequence of this attraction. Since the molecules hang on to each other, it is hard to get them moving and raise the temperature of water—a large number of calories are needed to heat the water. A metal has a low specific heat since the atoms are bonded together by electrons that range throughout the metal. It is easy to increase the motion of these free moving electrons, therefore easy to raise the temperature of the metal.

Thinking about that pizza again—we tried, but were unable to find the specific heat of cheese, tomato sauce, pepperoni, and crust. Often, when confronted with a complex physical analysis, it is best to simplify with a basic model. Assume the

Table 9.3. Specific heats of various materials.

Material	Specific heat (heat capacity) calories/gm – °C
water	1.0
air	0.241
aluminum	0.215
iron	0.108
carbon fiber [7]	0.359
copper	0.092

cheese, tomato sauce, and pepperoni to be mostly water and the specific heat is 1.0. For the crust, we approximate the dry crust as mostly air. The specific heat of air is 0.241. Therefore, it takes about four times as many calories of heat to raise the temperature of a gram of the cheese, etc as it does to raise the temperature of the same amount of crust. In addition, these juicy ingredients are more massive than the crust. When you take that bite, these tasty items deliver to your mouth much more heat than the dry crust.

Why does the rim of a bicycle get so warm after a long hard stop?

As a moving bicycle is brought to a stop, the sliding friction causes a significant amount of kinetic energy to be dissipated as heat between the brake pads and the rim. Typical rim materials, such as aluminum and carbon fiber, are like the pizza crust; they have a low specific heat and are easy to heat up.

Envision several scenarios involving heavy braking of a bicycle.

Scenario #1: rapid braking from a high speed

We can estimate the heat generated by looking at the kinetic energy change that occurs as a bike is brought to rest from an initial speed of about 35 mph. (For consistency with a subsequent calculation use a speed of 35.1 mph). Also, ignore the losses to air resistance; this energy serves to heat the air, and who worries about leaving behind a lot of hot air? Assume a worst-case scenario in which all of the bike/rider kinetic energy is converted into heat that is delivered to one wheel. In addition, the braking process is sufficiently fast that the spokes conduct little heat from the rim and there is insignificant energy loss to heat radiation from the rim.

A rider/bicycle system of mass 5.62 slugs traveling at 35.1 mph (51.5 ft/s) will possess the following kinetic energy:

$$KE = \frac{1}{2}mv^2 = \frac{1}{2} \times 5.62 \text{ slugs} \times (51.5 \text{ ft/s})^2 = 7453 \text{ ft-lb} = 2412 \text{ calories}$$

(1 calorie = 3.09 ft-lb, appendix B).

What would be the rim's final temperature if all of this heat were delivered to the rim of single bike rim? The specific heat of aluminum is 0.214 calories/gm; if the rim has a mass of 1000 gm, we can rearrange the specific heat expression equation (9.5) and solve for ΔT:

$$\Delta T = \frac{Q}{\text{c} \times \text{m}} = \frac{2412 \text{ calories}}{0.214 \dfrac{\text{calories}}{\text{gm} - {}^\circ\text{C}} \times 1000 \text{ gm}} = 11.3\ {}^\circ\text{C} = 52.3\ {}^\circ\text{F}$$

This increase in temperature, added to ambient temperature, will make the wheel warm to the touch but the rim and tire will be fine.

What occurs in a situation that involves more intense braking? Suppose we encounter a two-mile long, steep hill of 10% grade?

Scenario #2: riding the brakes down a hill

Our rider decides to cautiously proceed down the hill at a very slow speed; he 'rides the brakes' to hold the speed in check to only a few miles per hour. The result is that the bike/rider potential energy at the top of the hill is converted into thermal energy at the brake pads. To obtain a worst-case result, assume minimal losses to air resistance and other frictional forces.

A two-mile hill (10 560 ft) with a 10% grade ($\theta = 5.71°$) results in a vertical drop of 1051 ft; the loss of potential energy and equivalent amount of heat is:

$$\Delta PE = mgh = 5.62 \text{ slugs} \times 32.2\frac{\text{ft}}{\text{s}^2} \times 1051 \text{ ft} = 190\,000 \text{ ft-lb} = 61\,500 \text{ calories}$$

It is a bad idea but, if the braking is performed with one wheel, the change in temperature is:

$$\Delta T = \frac{Q}{\text{c} \times \text{m}} = \frac{61\,500 \text{ calories}}{0.214\frac{\text{calories}}{\text{gm} - °\text{C}} \times 1000 \text{ gm}} = 287\,°\text{C} = 549\,°\text{F}$$

The melting point of aluminum is about 1200 °F; there is no danger of melting the rim.

Perhaps our assumptions are extreme; however, sometime after a long hard descent, carefully touch your rim—it is definitely warm. Perhaps even more dramatic are the hot temperatures generated with disk brakes. Touching the rotors at the bottom of a long hill can produce an unpleasant burn. The reader is invited to do the calculation using the specific heat of steel and the mass of a rotor. An advantage of disk brakes is that the generated heat does warm up the rims.

Scenario #3: no braking until the bottom of a long hill

While this third scenario—flying down a long steep hill—might seem reckless to a novice rider, it results in minimum temperature impact on the rims. The reason is that the rider travels down the hill under a condition of terminal speed. From table 6.2 we see that, for our standard rider, the terminal speed on a 10% slope is 35.1 mph. If the rider brakes hard at the bottom of the hill, the rise in rim temperature will be the value calculated in scenario #1, that is a temperature increase of 52.3 °F

The system's potential energy at the top of the hill is converted into kinetic energy that, in turn, is lost to frictional heating of the air. Descending at high speed might well be 'safer' in regard to excessive heating of the rims, brake pads, and tires. Of course, the high speeds and hard braking at the bottom involves other safety considerations.

Why are carbon rims more susceptible than aluminum to overheating during a hard brake?

Carbon fiber wheels have a reputation for being more susceptible than metallic wheels to overheating. They have been determined to reach temperatures over 300 °F in hard braking [8]. The hot wheel temperatures create serious safety issues such as rim deformation and softening of tubular tire glue. When comparing the heating effects on different rim materials, several factors need to be considered:

- Specific heat of the materials. From table 9.3, we see that aluminum has a specific heat of 0.215; carbon fiber has a specific heat of 0.359. It is 'easier' to raise the temperature of aluminum when compared to the same mass of carbon.
- Mass of the rims. The advantage of carbon rims is that they can be made lighter than aluminum rims of the same strength. However, the lower mass carbon rims make them easier to heat up.
- Thermal conductivity. In table 9.2 of thermal conductivities, we see that aluminum has a conductivity of 204 Calories/h-m- °C and carbon fiber has a conductivity of 128 Calories/h-m- °C. In calculations on rim heating, we assumed that the heat is uniformly dispersed throughout the mass of the rim. With a rim material of lower conductivity, heat dispersal is less likely to occur. Aluminum does a better job than carbon fiber in conducting the heat throughout the entire rim. With carbon, heat generated near the pads will not be as readily dispersed throughout the rim's mass.
- Mechanical response to high temperatures. Metals, such as aluminum, will expand in a uniform manner to temperature changes. Carbon fiber materials, with their epoxy binders, have a transformation temperature much less than their melting point. Above this transformation temperature, the material begins to lose its rigid integrity and the substance becomes more plastic or rubbery in behavior.

References

[1] Lide D R (ed) 1990 *Handbook of Chemistry and Physics* 71st edn (Boca Raton, FL: CRC Press)

[2] Mosteller R D 1987 Simplified calculation of body surface area *New England J. Med.* **317** 1098

[3] Wegener C B 1997 Human adaption to hot environments www.dtic.mil/dtic/tr/fulltext/u2/a330520.pdf p 63

[4] Osczevski R and Bluestein M 2005 The New Wind chill equivalent temperature chart *Bull. Am. Meteorol. Soc.* **86** 1453–8

[5] Jendritzhy J *et al* The perceived temperature UTCI.org

[6] Lide D R (ed) 1990 *Handbook of Chemistry and Physics* 71st ed (Boca Raton, FL: CRC Press)

[7] A Packard, Reynolds Cycling, personal communication

[8] Shirley N 2011 Has their time really come—the where, why and how of carbon clinchers *Road Bike Action Magazine* July 2011 pp 70–4

Chapter 10

Rotational motion

All things come round to him who will but wait.

Longfellow, Students Tale

We now begin an exploration of the physics of rotary or circular motion. The application of a few principles is key to understanding many aspects of the bicycle's behavior:

- response of the foot pressing on pedals
- gearing systems
- energy of the spinning wheels
- turning of the handlebars
- balancing on a bicycle
- making turns
- self stability of the bicycle
- wheelies and headers

This approach to circular motion makes careful application of Newton's Laws of Motion. An understanding of inertia is critical in comprehending many situations of circular motion. These discussions on circular motion employ the methodology used in the study of linear motion; first, we look at the kinematics and later the dynamics of circular motion.

10.1 Kinematics of circular motion

Chapter 4 explored the kinematics of motion with concepts such as distance, speed, velocity, and acceleration. These motions are properly called 'linear' or 'translational' motions. In this section, we will see that there is often another type of motion associated with moving bodies. The motion is that of rotation; we will explore concepts such as angular distance, angular speed, angular velocity, and angular acceleration. No doubt, as the bicycle translates forward, there is quite a bit of rotational activity. The cranks spin, the gears turn, the wheels rotate, and, when

doi:10.1088/978-1-6817-4441-4ch10

making a turn, the entire system revolves about a point at the center of a circle. The concepts of rotational motion are best understood as analogs to linear motion.

Think of an object moving in a circle at a steady speed. It could be a car or a bike on a circular racetrack or even the tip of a hand on a clock. In figure 10.1, the clock face has only one hand—the minute hand. Suppose, initially, the minute hand points directly upward at twelve o'clock and ten minutes later the hand is aiming at two o'clock. The hand has swept out an angle θ of 60° in the clockwise sense. We also envision the tip of hand as sweeping out an arc of length s.

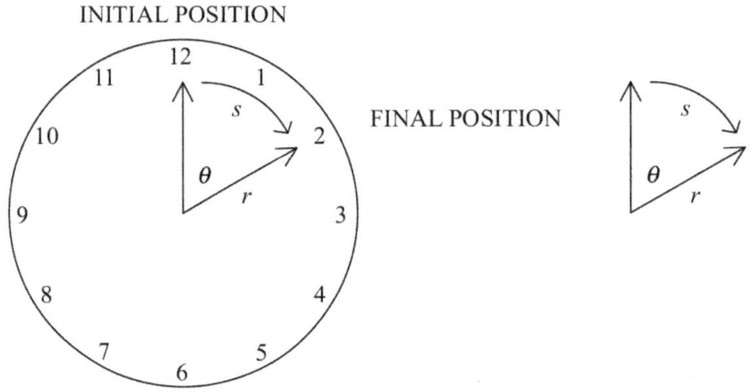

Figure 10.1. Clock face.

This simple situation introduces a new way to describe angles; we define it to be the ratio of the arc length s to the length of the clock hand r:

$$\theta = \frac{s}{r} \tag{10.1}$$

When an angle is specified in this manner, the units are called *radians*. To get a feeling for these units of radians, we put some numbers in the example. Suppose the length of the second hand is six inches, the circumference of the circle will be 37.7 inch. The arc length from 12 o'clock to 2 o'clock positions will be one sixth of the circumference or 6.28 inch. Thus for the angle, we obtain:

$$\theta = \frac{s}{r} = \frac{6.28 \text{ inch}}{6 \text{ inch}} = 1.05 \text{ radians}$$

Instead of saying the angle is 60°, we could say that the angle is 1.05 radians! Is there any advantage to the use of radians to measure angles? Yes, but a bit of patience is necessary to see the advantage develop. If the arc is the full circumference, the arc length is $2\pi r$ and the angle of a full circle of any size is:

$$\theta = \frac{s}{r} = \frac{2\pi r}{r} = 2\pi \text{ radians} = 6.28 \text{ radians}$$

The angle for a full circle is 360° or 2π radians. The ratio gives us a useful conversion:

$$\frac{360^\circ}{2\pi \text{ radian}} = 57.3^\circ \text{ per radian}$$

This example can be extended into rotational motion by considering further the minute hand as it moves a distance s from 12 o'clock to 2 o'clock in a time t (normally ten minutes). Using the concepts of chapter 4, we write the linear speed as:

$$v = \frac{\text{distance}}{\text{time}} = \frac{s}{t} \tag{10.2}$$

Dividing both sides by the length of the hand r:

$$\frac{v}{r} = \frac{s}{rt} = \frac{s/r}{t} = \frac{\theta}{t} \tag{10.3}$$

The right side is called the *angular speed*, usually represented with the Greek letter ω (lower case omega).

$$\omega = \frac{\theta}{t} \tag{10.4}$$

Another useful expression connecting the linear speed with the angular speed is obtained by combining the two previous equations:

$$\frac{v}{r} = \omega; \quad \text{or} \quad v = \omega r \tag{10.5}$$

If our clock is running fine, it traverses the 1.05 radian angle in ten minutes; the angular speed in radians per second is:

$$\omega = \frac{1.05 \text{ radians}}{600 \text{ s}} = 0.00175 \text{ radians/s}$$

As an example of the application of these angular concepts to the bicycle—suppose we ask how many rotations will a wheel make if a cyclist travels 3000 miles? It is important to use a consistent set of units—get all the distances in feet:

$$s = 3000 \text{ mile} \times \frac{5280 \text{ ft}}{\text{mile}} = 15\,840\,000 \text{ ft}$$

A bike, with wheel diameter of 27 inches, has a radius of 1.125 ft; the angle traversed in the 3000 mile trip is:

$$\theta = \frac{s}{r} = \frac{15\,840\,000 \text{ ft}}{1.125 \text{ ft}} = 14\,080\,000 \text{ radians}$$

since one revolution of the wheel is 2π radians:

$$\text{number of revolutions} = \frac{14\,080\,000 \text{ radians}}{2\pi} = 2\,242\,038 \text{ revolutions}$$

A lot of spinning!

10.1.1 Angular velocity

These new angular measurements introduce another way to think of motion. Since the tips of the clock's hands are executing angular motion, they have an *angular speed* and the associated vector quantity, the *angular velocity*. The direction of the angular velocity will initially seem strange—it is given by the *right hand rule*: To obtain the direction of an angular velocity vector, stick the thumb out and curl the fingers of the right hand along the sense of the rotation (clockwise or counter-clockwise); the angular velocity vector points in the same direction as the thumb. Figure 10.2 shows a toy car traveling counterclockwise; its angular velocity is pointing upward. The hands of a wall clock have an angular velocity that points into the wall. A forward moving bicycle wheel will have an angular velocity that points along the hub toward the left.

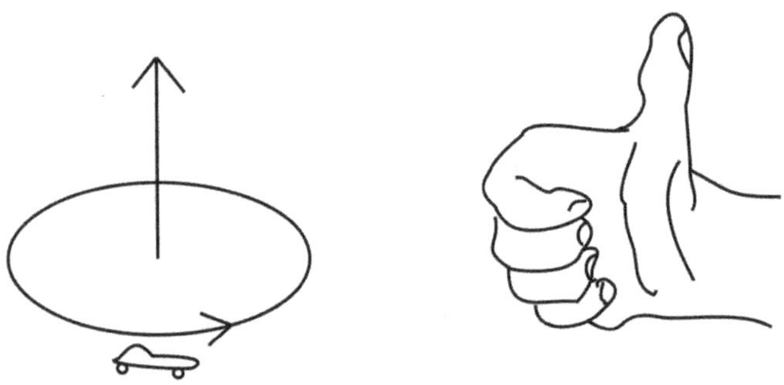

Figure 10.2. Right hand rule.

10.1.2 Angular acceleration

In a clock that is operating properly, the hands will run at constant angular speeds/velocity. However, what if the clock is started up but the minute hand requires a few seconds to reach its normal speed of 0.00175 rad/s? The hand is undergoing an *angular acceleration*; it is gaining rotational speed. The angular acceleration α is defined as the change in angular speed with time.

$$\alpha = \frac{\omega_f - \omega_i}{t} \tag{10.6}$$

If the hand starts from rest ω_i and needs three seconds to reach normal speed ω_f, the angular acceleration is:

$$\alpha = \frac{0.00175 \text{ radians/s} - 0}{3 \text{ s}} = 0.000583 \text{ radians/s}^2$$

To gain a better understanding of these angular concepts, consider another example of circular motion—a cyclist riding around a circular path. If she rides at a steady linear speed v_i of 10 mph (14.7 ft/s) in a circle of radius 50 ft, her initial angular speed is written:

$$\omega_i = \frac{v_i}{r} = \frac{14.7 \text{ ft/s}}{50 \text{ ft}} = 0.294 \text{ radians/s}$$

While riding at these constant linear and angular speeds, she is only pedaling hard enough to overcome the resistive forces described in chapter 5. Table 5.2 shows that, at 10 mph, she must press on the pedals hard enough to balance a total resistive force of 2.39 lb. After riding for a time at 10 mph, she decides to add an extra kick and one half second later her speed v_f is 11 mph (16.1 ft/s); thus her final angular speed is

$$\omega_f = \frac{v_f}{r} = \frac{16.1 \text{ ft/s}}{50 \text{ ft}} = 0.322 \text{ radians/s}$$

Her linear acceleration is:

$$a = \frac{v_f - v_i}{t} = \frac{16.1 \text{ ft/s} - 14.7 \text{ ft/s}}{0.5 \text{ s}} = 2.80 \text{ ft/s}^2$$

and her angular acceleration is

$$\alpha = \frac{\omega_f - \omega_i}{t} = \frac{0.322 \text{ radians/s} - 0.294 \text{ radians/s}}{0.5 \text{ s}} = 0.056 \text{ radians/s}^2$$

If her acceleration is constant, her average linear and angular speeds are:

$$v_{ave} = \frac{v_f + v_i}{2} = \frac{16.1 \text{ ft/s} + 14.7 \text{ ft/s}}{2} = 15.4 \text{ ft/s}$$

$$\omega_{ave} = \frac{\omega_f + \omega_i}{t} = \frac{0.322 \text{ radians/s} + 0.294 \text{ radians/s}}{2} = 0.308 \text{ radians/s}$$

In the one half second of acceleration, she will traverse an angular distance of

$$\theta = \omega_{ave} \times t = 0.308 \text{ radians/s} \times 0.5 \text{ s} = 0.154 \text{ radians} \times \frac{57.3^\circ}{1.0 \text{ radian}} = 8.82^\circ$$

Another useful expression for the angular acceleration may be obtained by returning to equation (10.6) and adding a few extra steps:

$$\alpha = \frac{\omega_f - \omega_i}{t} = \frac{\Delta\omega}{t} = \frac{\Delta v/r}{t}$$

since $\frac{\Delta v}{t}$ is the linear acceleration a, we obtain a connection between the angular acceleration and the linear acceleration:

$$\alpha = a/r \tag{10.7}$$

10.2 Dynamics of circular motion

The primary advantage of these circular motion concepts occurs when we explore *rotational dynamics*—the role of forces in producing angular accelerations. From Newton's Laws of Motion in chapter 5, we learned that changes in linear motion of a body, i.e. accelerations, are caused by net external forces. In the realm of rotational motion, the rotational analogue to a force is a physical concept called *torque*. Changes in rotational motion, i.e. rotational accelerations, are caused by net external torques.

To produce acceleration the cyclist, riding in the circle, must press hard enough on the pedal to create a net external force that is greater than the resistive forces. Newton's Second Law gives the value of the force; to produce an acceleration of 2.80 ft/s^2 on a bike rider system of 5.62 slugs, the net external force in the forward direction is:

$$F = ma = 5.62 \text{ slugs} \times 2.80 \text{ ft/s}^2 = 15.7 \text{ lb}$$

Since the rider's acceleration is tangent to the circle, this force is also tangent to the circle. These vectors are illustrated in the figure 10.3 below.

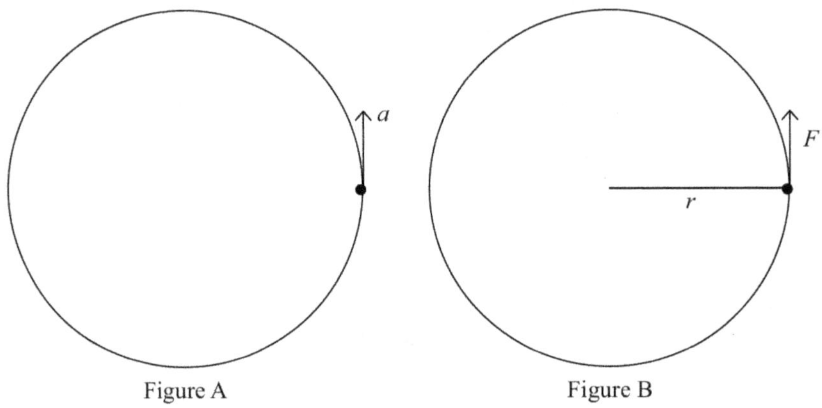

Figure 10.3. (A,B) Circular motion.

Shown in the right hand drawing (figure 10.3B) is a line representing the circle's radius. This situation affords an opportunity to introduce the new physical quantity *torque*. The radial distance r is called the *lever arm*; the product of the lever arm and the force is the torque. Traditionally, the Greek letter τ (pronounced 'tau') is the mathematical symbol for torque, thus

$$\text{torque} = \text{lever arm} \times \text{force}$$

in letters

$$\tau = r \times F \tag{10.8}$$

(The next chapter explores further aspects of torque; the more descriptive d$\perp$ is used as the symbol for the lever arm.)

To obtain the torque on the cyclist accelerating about the 50 ft circle, we have:

$$\text{torque} = \text{lever arm} \times \text{force} = 50\text{ ft} \times 15.7\text{ lb} = 785\text{ lb-ft}$$

Since the ft-lb is the unit of work and energy in the U.S. Customary system, the units of torque are typically written as lb-ft; another common unit for torque measurement is the lb-inch.

Returning to Newton's Second Law:

$$F = ma$$

and multiplying both sides by r:

$$r \times F = r \times m\, a$$

The left side is torque; on the right side, we use equation (10.7) to substitute for the acceleration $a = \alpha\, r$

$$\tau = rm\,\alpha r = mr^2\,\alpha = I\alpha$$

where we have replaced mr^2 with a physical expression called the *moment of inertia*. The letter I is the symbol for moment of inertia.

$$I = mr^2 \tag{10.9}$$

The units of moment of inertia are 'slugs-ft^2'; they have no special name. The resulting expression is written:

$$\tau = I\alpha \quad \text{or} \quad \alpha = \frac{\tau}{I}$$

and is the rotational analogue to Newton's Second Law

$$F = ma \quad \text{or} \quad a = \frac{F}{m}$$

Angular acceleration is the rotational analogue to linear acceleration and moment of inertia is the rotational analogue to mass; torque is the rotational analogue to force. Just as we say forces cause linear accelerations, we can say torques cause angular accelerations. Masses require forces to accelerate; moments of inertia require torques to change their rotation. Individual forces are summed to obtain a net external force, individual torques are summed to obtain net external torque.

Each of the linear concepts of motion has a rotational analogue summarized in the table 10.1.

Although the basic idea of torque has been introduced in this section, it has so many nuances and applications to the bicycle that we devote the entire next chapter to the physics of torque.

Before leaving this section, it is worthwhile to point out that, in addition to the resistive backward forces and the net forward force discussed above, there is another unbalanced force on the bike/rider system. This force acts perpendicular to the body's velocity along the radius directed toward the center of the circle. The force, called the *centripetal force*, will be the topic of much discussion in chapter 12.

Table 10.1. Comparison of linear and rotational motion parameters.

Linear motion	Rotational motion
distance: s	angle θ (= s/r)
speed: v	angular speed: ω (= v/r)
acceleration: a	angular acceleration: α (= a/r)
force: F	torque: τ (= $r \times F$)
inertia (linear): m	inertia (rotational): I

10.3 Rotational kinetic energy

This section explores further aspects of rotational motion. A simple situation of a stone on a rope being whirled in a horizontal circle is a good example (figure 10.4A). The view from above is also shown (figure 10.4B).

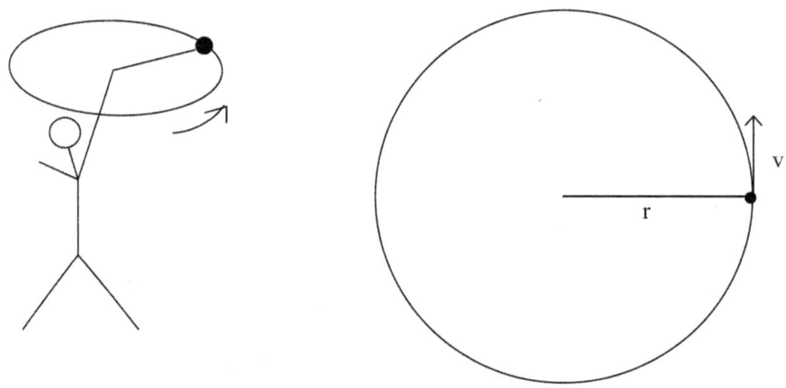

Figure A Figure B

Figure 10.4. (A,B) Whirling a rock.

If the rock has a mass m and is traveling at a constant linear speed v, the rock's kinetic energy is:

$$KE = \frac{1}{2}mv^2 \tag{10.10}$$

With a circle of radius r, the linear speed is related to the angular speed: $v = \omega r$

The kinetic energy becomes:

$$KE = \frac{1}{2}m\omega^2 r^2$$

Grouping the mass and radius terms:

$$KE = \frac{1}{2}(mr^2)\omega^2$$

The term in parentheses is the moment of inertia: $I = mr^2$

The whirling rock's kinetic energy is now considered as a rotational kinetic energy and is written:

$$KE_{\text{rot}} = \frac{1}{2}I\omega^2 \tag{10.11}$$

This expression is the rotational analogue to linear kinetic energy equation (10.10).

Putting some numbers into this expression: suppose a rock of mass 0.0311 slugs (one pound) is whirled at three revolutions per second in a circle of 2.5 ft radius.

The angular speed will be:

$$\omega = \frac{3\text{ revolution}}{\text{s}} \times \frac{2\pi\text{ radians}}{\text{revolution}} = 18.8\text{ radians/s}$$

The rock's moment of inertia is:

$$I = mr^2 = 0.0311\text{ slugs} \times (2.5\text{ ft})^2 = 0.194\text{ slugs-ft}^2$$

For the rotational kinetic energy:

$$KE_{\text{rot}} = \frac{1}{2}I\omega^2 = \frac{1}{2} \times 0.194\text{ slugs-ft}^2 \times (18.8\text{ radians/s})^2 = 34.3\text{ ft-lb}$$

The reader might wish to verify that this result is the same as obtained by computing the linear speed based on the circumference of the circle and the time of revolution. The real value of using rotational concepts comes when we are dealing with non-point masses such as hoops, disks, rods etc.

10.4 Moment of inertia of non-point masses

Our examples thus far have illustrated moving objects considered as point masses—that is, all of the object's mass was concentrated in a size small compared to other distances in the problem. For example, the diameter of the rock was small compared to the length of the rope.

> Just as mass is a measure of a body's inertial tendency to resist changes in linear motion, the moment of inertia is a measure of the body's tendency to resist changes in rotational motion.

The fact that the moment of inertia contains the square of the radius tells us that the location or distribution of the mass has a major impact on the moment of inertia. This effect is easy to demonstrate on a kitchen refrigerator door with large shelves that can hold gallon size bottles of milk. The radius of rotation is determined by the distance of the milk bottles from the hinge axis. When the bottles are located close to the hinges, the door is easy to swing; whereas, moving the same bottles out to the edge of the door makes the door noticeably harder to swing open and close. Swinging the door provides an intuitive feeling for the moment of inertia.

With rotating bodies, not only is the mass of the body important, but also the *distribution of the mass* about the axis of rotation. In general, the moments of inertia for a body of a specific shape can be obtained using calculus; however, for many simple shapes it is possible to use intuition to obtain an object's moment of inertia. We offer a few below.

Hoop—such as a bicycle rim—if all of the mass is concentrated at a fixed distance from a central axis rotation, the moment of inertia is the same as a point mass, that is:

$$I_{\text{hoop}} = mr^2 \tag{10.12}$$

Solid disk—such as a coin rotated about central axis perpendicular to the face—if the mass were uniformly distributed, we could envision it as a series of concentric rings and the moment of inertia would be one half that of a ring of the same mass:

$$I_{\text{disk}} = \frac{1}{2}mr^2 \tag{10.13}$$

The same formula would apply for a solid cylinder rotating about its central axis.

Long thin rod—such as a bicycle spoke—if it has a length l and a mass m and is rotated about an axis on one end, it will have a moment of inertia less than one half of the point mass:

$$I_{\text{rod}} = \frac{1}{3}ml^2 \tag{10.14}$$

The reader is advised to talk to her local calculus teacher for exact proofs of the above. In the next section, several of these shapes are combined to obtain the moment of inertia for a bicycle wheel.

10.5 Moment of inertia and rotational kinetic energy of bicycle wheel

Objects of complex shapes, such as a bicycle wheel, can be divided into a few simple components. In the wheel, the mass is found in three locations: the periphery (rim, tire, and tube), the spokes, and the hub. The rim, tire, and tube are hoops; the spokes are long thin rods; the hub may be approximated as a solid disk. Since all parts are rotating about the wheel's center axis, we can sum the individual moments (figure 10.5).

Since riders of high end bicycles are very conscious of their bike's weight, there are published values of the wheel components[1]. Depending on the price point, there is considerable variation in masses, the more expensive being the least massive. The table below lists typical values for mid-priced wheels (Table 10.2).

[1] www.Wheelbuilder.com is an excellent source of data.

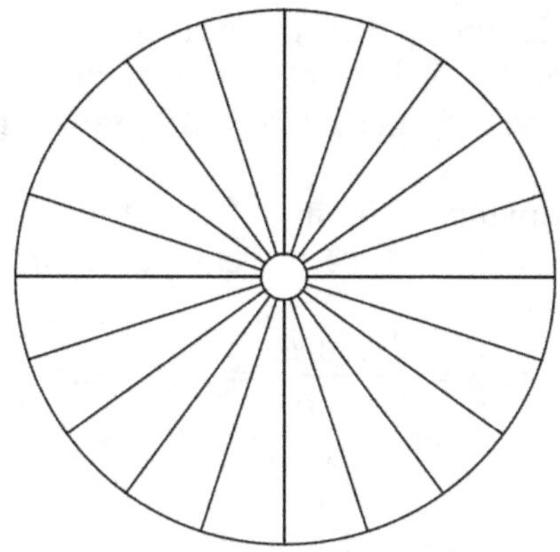

Figure 10.5. Wheel, spokes, and hub.

Table 10.2. Wheel components mass and geometry.

Component	Mass in slugs	Geometry
rim	0.035	hoop
tire	0.014	hoop
tube	0.007	hoop
spokes—22 count	0.008	rod
hub	0.010	cylinder

Rim, tire, tube. Since the rim, tire, and tube share the same geometry, we can add their masses and consider their radii to be 1.13 ft (27 in diameter wheel).

$$I_{\text{rim tire tube}} = I_{\text{hoop}} = m\,r^2 = (0.035 \text{ slugs} + 0.014 \text{ slugs} + 0.007 \text{ slugs})(1.13 \text{ ft})^2$$
$$= 0.0715 \text{ slugs-ft}^2$$

Spokes. The spokes are long narrow rods, rotated about one end; ignore the distance from the inner end of the spoke to the center of the wheel's rotation. A typical spoke length is about 0.820 ft.

$$I_{\text{spokes}} = I_{\text{rod}} = \frac{1}{3}ml^2 = \frac{1}{3}(0.008 \text{ slugs})(0.820 \text{ ft})^2 = 0.00179 \text{ slugs-ft}^2$$

Hub. The construction is complex with bearings, flanges, cones, etc. The good news is that all parts of the hub are very close to the axis of rotation and will have minimal impact on the wheel's overall moment of inertia. A simplification is obtained by using the geometry of a solid cylinder of radius 0.08 ft.

$$I_{\text{hub}} = I_{\text{disk}} = \frac{1}{2}mr^2 = \frac{1}{2}(0.010 \text{ slugs})(0.08 \text{ ft})^2 = 0.00003 \text{ slugs-ft}^2$$

Adding the above three results to get a value for the overall wheel's moment of inertia:

$$I_{\text{wheel}} = I_{\text{rim tire tube}} + I_{\text{spokes}} + I_{\text{hub}}$$
$$I_{\text{wheel}} = 0.0715 + 0.00179 + 0.00003 = 0.0733 \text{ slugs-ft}^2$$

The calculation reveals that the spokes and hub contribute only a small amount to the wheel's moment of inertia. A careful reader might inquire about the air in the tire; she is invited to use the air's mass (about 10 grams) in a similar fashion to the rim, tube and tire.

Using the wheel's moment of inertia, the rotational kinetic energy may be calculated. If the wheel is on a bicycle traveling at 20 mph (29.3 ft/s), the angular speed will be:

$$\omega = \frac{v}{r} = \frac{29.3 \text{ ft/s}}{1.13 \text{ ft}} = 25.9 \text{ radians/s}$$

The rotational kinetic energy of the wheel is:

$$KE_{\text{rot}} = \frac{1}{2}I\omega^2 = \frac{1}{2}(0.0733 \text{ slugs-ft}^2)(25.9 \text{ radians/s})^2 = 24.6 \text{ ft-lb}$$

While the front and rear wheels will be slightly different due to spoke counts and hub construction, doubling the above value gives a good approximation for the rotational kinetic energy of both wheels.

How does this number compare to the translational (linear) kinetic energy of the total system consisting of the bike and rider? For a 161 lb rider on a 20 lb bike, including wheels, (5.62 slugs), the translational kinetic energy at 20 mph is:

$$KE_{\text{translation}} = \frac{1}{2}mv^2 = \frac{1}{2}(5.62 \text{ slugs})(29.3 \text{ ft/s})^2 = 2412 \text{ ft-lb}$$

$$KE_{\text{total}} = KE_{\text{translation}} + 2 \times KE_{\text{rotating wheels}} = (2412 \text{ ft-lb}) + 2(24.6 \text{ ft-lb})$$
$$= 2461 \text{ ft-lb}$$

Note that the wheels possess two forms of kinetic energy—translational kinetic energy associated with the system's linear speed and rotational kinetic energy due to the spinning of the wheels.

In this example, the rotational contribution to the total kinetic energy is a small fraction, approximately 2%, of the total. For a 'cruiser' type bike with massive wheels and tires, the rotational kinetic energy would be several times larger, but nevertheless small compared to translational kinetic energy.

10.6 Angular momentum

In this section, we continue to explore rotational analogues to linear motion. In chapter 7 we saw the concept of momentum. For the rock of mass m traveling at a speed v, the momentum p is defined as:

$$p = mv \tag{10.15}$$

Making the following substitutions for the rotational equivalents:

$$m = \frac{I}{r^2} \quad \text{and} \quad v = \omega r$$

We obtain:

$$p = \frac{I}{r^2}(\omega r) = \frac{I\omega}{r}$$
$$pr = I\omega$$

The right side is called the angular momentum L:

$$L = I\omega \qquad (10.16)$$

The angular momentum is the rotational analogue to linear momentum equation (10.15).

The whirling rock of section 10.3 may be used for an angular momentum calculation. The rock of mass 0.0311 slugs is whirled at 3.0 revolutions per second in a circle of 2.5 ft radius. The moment of inertia is 0.194 slugs-ft^2 and the angular speed is 18.84 rad/s. We obtain for the angular momentum:

$$L = I\omega = 0.194 \text{ slugs-ft}^2 \times 18.84 \text{ radians/s} = 3.65\frac{\text{slugs-ft}^2}{\text{s}}$$

What is the value of this new quantity called angular momentum? As we saw in chapter 7, linear momentum is changed by net external forces; angular momentum is changed by net external torques. In the absence of these net external forces and torques, the momenta are constant. A formal statement of the law of conservation of angular momentum is:

In the absence of net external torques, the angular momentum of a system is conserved

With no external torques, the initial angular momentum must equal the final angular momentum:

$$L_i = L_f$$
$$I_i\omega_i = I_f\omega_f$$

This conservation of angular momentum is easy to demonstrate with the whirling rock. As the rock is being whirled by the right hand, use the left hand to pull downward on the rope's loose end and therefore, reduce the size of the rock's orbit. The rope tension is radially directed and produces no torque. As the radius of the orbit shortens, perhaps from 2.5 ft to 1.25 ft, and the moment of inertia is reduced

$$I_f = mr^2 = 0.0311 \text{ slugs } \times (1.25 \text{ ft})^2 = 0.0486 \text{ slugs-ft}^2$$

The final angular speed is now:

$$\omega_f = \frac{I_i\omega_i}{I_f} = \frac{\frac{3.65 \text{ slugs-ft}^2}{\text{s}}}{0.0486 \text{ slugs-ft}^2} = 75.1 \text{ radians/s} \times \frac{1 \text{ revolution}}{2\pi \text{ radians}} = 12.0\frac{\text{revolutions}}{\text{s}}$$

As the rope is shortened, the rock spins much faster. Conservation of angular momentum is gracefully demonstrated by a figure skater when she begins a spin with limbs extended. As she draws them toward her body, her moment of inertia is reduced and she ends up spinning very rapidly.

10.6.1 Direction of angular momentum

Angular momentum is a vector quantity whose direction is as important as its magnitude. Since the direction of angular speed is given by the right hand curled finger rule with the velocity vector pointing along the thumb, the direction of the angular momentum is given by the same right hand rule. If a cyclist curls the fingers of her right hand in the sense of the wheels' rotation, she will see that the angular velocity and angular momentum of the wheels are vectors that point to the left.

Conservation of angular momentum applies to both magnitude and direction. This conservation of the angular momentum gives a rotating object stability along the axis of rotation. Angular momentum is a vector and wants to remain pointing in a fixed direction. A quarterback will impart a spin to a pass in order to minimize the football's wobble. Other examples are spinning tops and gyroscopes.

10.7 Role of angular momentum in a bicycle

Does the gyroscopic action of the rotating wheel play a role in balancing a bike?

> Angular momentum is a very important physical concept in any discussion of bicycle physics. It is significant because it is commonly misunderstood and applied in a flawed manner. The angular momentum of the rotating wheels is erroneously invoked as the cause of the bike's stability. The analogy is drawn to a spinning top or gyroscope.

Why isn't a bicycle's balance caused by the gyroscopic action of two wheels rolling along the ground?

Sometimes, a bicycle wheel is compared to a simple hoop; give the hoop a toss and, as long as it is rolling, it remains upright. As soon as the hoop loses speed, and therefore angular momentum, the hoop falls over. Is this not the same mechanism keeping the moving bicycle up? The answer is no. The magnitudes of angular momentum for a rolling hoop versus a falling hoop tell the story.

A hoop of any size can be used; it is only important to give it a good roll. Consider a simple hoop (figure 10.6A) with all of its mass along the circumference; if the hoop's mass is 0.3 slugs and radius is 1.5 ft, the moment of inertia is

$$I_{\text{hoop}} = mr^2 = 0.5 \text{ slugs} \times (1.5 \text{ ft})^2 = 1.13 \text{ slugs-ft}^2$$

When traveling at 20 mph (29.3 ft/s), the angular momentum is:

$$L_{\text{rolling hoop}} = I\omega = I\frac{v}{r} = (1.13 \text{ slugs-ft}^2) \times \frac{29.3 \text{ ft/s}}{1.5 \text{ ft}} = 22.1\frac{\text{slugs-ft}^2}{\text{s}}$$

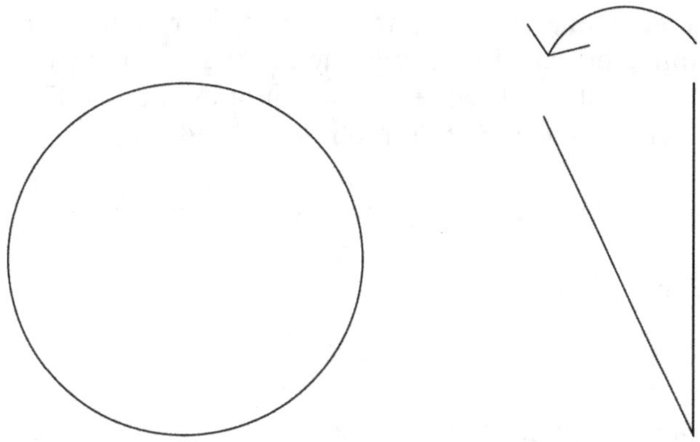

VIEW FROM THE SIDE — Figure A

VIEW FROM BEHIND — Figure B

Figure 10.6. (A,B) Hoop.

This angular momentum direction is given by the right hand curled finger rule, along the axis of the hoop.

Now think about the angular momentum of a hoop falling over. Viewing the hoop from behind, it appears as a long thin rod (figure 10.6B).

If the hoop is falling to the left, we approximate it as a rod of length 3 ft rotating about one end:

$$I_{\text{rod}} = \frac{1}{3}ml^2 = \frac{1}{3}(0.5 \text{ slugs}) \times (3 \text{ ft})^2 = 1.5 \text{ slugs-ft}^2$$

As the hoop falls to one side, the angular speed is not constant since the force of gravity pulls down with an ever-increasing torque. We estimate the average angular speed by supposing the hoop falls over in one second, traversing an arc of 90° (pi/2 radians):

$$\omega = \frac{\theta}{t} = \frac{1.57 \text{ radians}}{1 \text{ s}}$$

The angular momentum of the falling hoop will be:

$$L_{\text{falling hoop}} = I_{\text{rod}}\,\omega = (1.5 \text{ slugs-ft}^2) \times (1.57 \text{ radians/s}) = 2.36\frac{\text{slugs-ft}^2}{\text{s}}$$

Considering the ratio:

$$\frac{L_{\text{rolling hoop}}}{L_{\text{falling hoop}}} = \frac{22.1}{2.36} = 9.36$$

The rolling hoop possesses an angular momentum much larger than the same hoop falling to the side. The tendency for the rolling hoop's angular momentum to remain constant affords stability, resisting the tendency of the hoop to fall to the side. The hoop readily falls when it slows and loses it rolling angular momentum.

We now apply the same analysis comparing the angular momentum of a moving bicycle to that of a falling bicycle. First, compute the angular momentum of the bike wheel discussed in section 10.5. The wheel's moment of inertia is 0.0733 slugs-ft^2 and the angular velocity when traveling at 20 mph is 25.9 rad/s.

$$L = I\omega = (0.0733 \text{ slugs-ft}^2) \times (25.9 \text{ radians/s}) = 1.90\frac{\text{slugs-ft}^2}{\text{s}}$$

Doubling this value for our two-wheeler:

$$L_{\text{two wheels}} = 3.80\frac{\text{slugs-ft}^2}{\text{s}}$$

Now consider the angular momentum of a cyclist falling to one side. Getting the angular momentum requires a calculation of his moment of inertia and speed of falling. We will try for some type of estimate.

A crouched rider on a racing bike has a center of mass somewhere near his midsection, approximately 3.80 ft above the ground. To get a value for the system's moment of inertia, we treat the bike and rider as a point mass and write for the moment of inertia for the 181 lb (5.62 slugs) system:

$$I_{\text{bike + rider}} = mr^2 = 5.62 \text{ slugs} \times (3.80 \text{ ft})^2 = 81.2 \text{ slugs-ft}^2$$

Notice, the rider and bike system has a much larger moment of inertia than that of the two wheels. This is a significant clue that the rotation of the wheels is not going to keep the rider upright by some type of gyroscopic effect. To calculate angular momentum of the falling rider, suppose he falls over in one second (it just seems like an hour when happening), the average angular speed is then:

$$\omega = \frac{\Delta\theta}{\Delta t} = \frac{1.57 \text{ radians}}{1.0 \text{ s}} = 1.57\text{radians/s}$$

Thus, the angular momentum of the falling rider/bike:

$$L_{\text{bike + rider}} = I\omega = (81.2 \text{ slugs-ft}^2)(1.57\text{radians/s}) = 127\frac{\text{kg-m}^2}{\text{s}}$$

Comparing the ratio of the angular momenta of the rotating wheels to the falling system:

$$\frac{L_{\text{two wheels}}}{L_{\text{bike + rider}}} = \frac{3.80}{127} = 0.0299$$

The angular momentum of the bike's rotating wheels is but a small fraction (about 3%) of the bike/rider system's angular momentum as it is falling sideways. The rolling wheels' angular momentum is far too insignificant to stabilize the bike from a sideways fall.

A second argument can be made against angular momentum being the stabilizing factor for a bicycle. In a comparison between a modern racing bike with thin lightweight wheels and a heavy fat wheeled cruiser machine, the racing bike is extremely stable and no harder to balance than the cruiser. While it is true that racing bikes have a nervous, skittish personality, it is a consequence of a short wheelbase and lightweight construction—not a lack of angular momentum in the wheels.

Finally, one more argument—perhaps the most compelling—can be raised against the wheel's angular momentum being the stabilizing factor in a bicycle. If the front wheel is locked such that the handlebars cannot turn, the bike will immediately fall over, regardless of how fast the rider spins the wheels. Chapter 12 examines the causes for a bicycle's stability.

Chapter 11

Torque—applications to the bicycle

In the last chapter, we explored rotational motion concepts such as angular speed, angular velocity, and angular acceleration. Rotational motion was viewed as analogous to linear motion. The concept of torque was introduced as the rotational analogue of force; just as net external forces cause accelerations in linear motion, net external torques produce rotational accelerations in circular motion.

In this chapter, we consider in detail common occurrences of torque, especially torque's numerous applications to the bicycle. A note—in the world of engineering, torques are called moments of force.

11.1 Basic physics of torque

Torque, first seen in the previous chapter, is an extremely important physical concept that has many nuances and important applications to the human body while on and off a bicycle. Whether we are pedaling at constant speed, accelerating, decelerating, or turning, the bicycle is a marvelous application of torque physics. Torques will give us insight into the bicycle's gearing and braking efficiencies. Torques will help us understand stunts such as wheelies and the disasters of headers. We will see why the front brake is more effective than the rear brake in quick stops.

Consider a familiar event: the opening of a can of paint by prying off the lid with a long screwdriver. The usual method is to catch the end of the blade under the edge of the lid and push the handle downward (figure 11.1A).

The downward force exerted by the painter results in a rotation of the screwdriver about the outside edge of the can. The point of rotation is called the *fulcrum*—sometimes called the pivot point (figure 11.1B). This common act is a nice illustration of the physical quantity *torque*. Torques cause rotations—in our example, it is the rotation of the paint can lid. Experience shows that the length of the handle and the hand's position on the screwdriver handle will affect the ease in opening the can. Applying the force at point a will open the can much easier than the same downward force applied at point b. There are at least two important factors—the magnitude of

doi:10.1088/978-1-6817-4441-4ch11

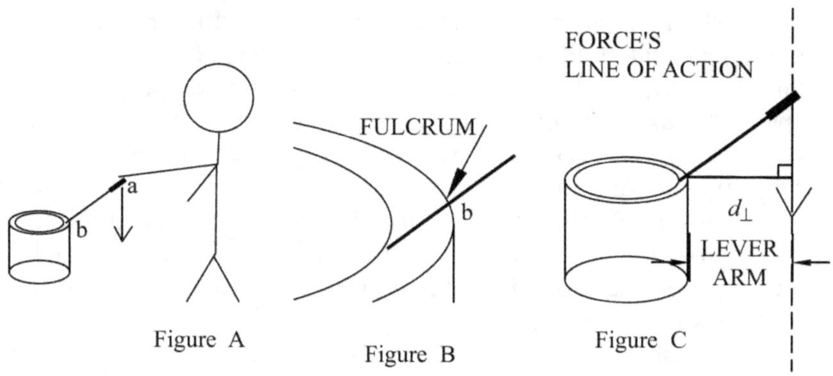

Figure 11.1. (A,B,C) Opening a can of paint.

the force and the distance of the force from the fulcrum. Further reflection will reveal that there is one more factor—the angle at which the force is applied. For instance, if we apply the force parallel to the screwdriver handle, the lid will not pop off. Torque involves three factors: the magnitude of the force, the distance from the force's point of application to the fulcrum, and the direction of the force. The direction of force is called the line of action.

With these concepts of the force's line of action and the lever arm drawn between the fulcrum and then perpendicular to the line of action, the magnitude of the torque is defined as:

$$\text{torque} = \text{lever arm} \times \text{force}$$

It is traditional to use the Greek letter τ (tau) as the symbol for a torque, with $d\perp$ for the lever arm and F for the force. The above expression becomes:

$$\tau = d\perp \times F \tag{11.1}$$

Note: in the previous chapter, when we introduced torques in the context of a body moving in a complete circle, we used the radius of the circle r as the variable for the lever arm.

11.2 Rotational equilibrium

Torque is the rotational analogue to force. *Just as net external forces are the cause of linear accelerations, net external torques are the cause of rotational accelerations.*

Painting might not be much fun; let us turn to another example that is almost as much fun as a bike ride. Think about two kids in a playground on a seesaw. Envision an old-fashioned seesaw consisting of a long board balanced on a pole. Normally, there is some way to shift the board along its length so kids of different weights can balance the seesaw. Consider that we have two youngsters—a 60 lb girl and her 30 lb little brother. How might they balance on a seesaw board that is 12 ft long? Children learn quickly that the small child must have more of the board on his side of the balance point. In fact, using intuition and a bit of practice, the kids will realize that, since the girl is twice the weight of the boy, the boy must be twice the distance from the fulcrum. Each child will produce the same torque and the net torque is zero—hence, the seesaw is balanced: it experiences zero net torque.

This simple playground experience introduces a new physical principle—the principle of *rotational equilibrium*. It is considered a rotational analogue to Newton's First Law. The First Law stated that an object at rest or moving at a constant velocity will continue to do so unless acted upon by a net external force. The condition was that of *translational equilibrium*. When looking at the equilibrium of forces, we determine if the horizontal forces to the left are equal to the horizontal forces to the right. Similar conditions apply to vertical forces up and down and to forces in and out of the paper.

For the situation of *rotational equilibrium*, where there are no changes in rotational velocities, the requirement is to have *no net torque*, i.e. a balance of torques. When looking at rotations, the nomenclature of left/right or up/down is not appropriate—that rule of 'righty tightie/lefty loosie' was never very clear. It is much better to describe a *rotation about the fulcrum by its sense of clockwise or counterclockwise*. In a two dimensional situation, such as the seesaw, an object will be in rotational equilibrium if the clockwise torques are equal to the counterclockwise torques.

11.3 Mechanical advantage

The concepts of torque and rotational equilibrium can be developed further with another example employing lever arms. Consider another common application, using a long bar to lift a heavy boulder. Suppose a girl is trying to raise a 320 lb rock by wedging a strong 9 ft bar over a fulcrum and pushing down on the end opposite from the rock. In figure 11.2, the weight of the rock acts at one end and the girl pushes downward at the other end.

In figure 11.2, the girl is shown using the fulcrum in two different positions. In figure 11.2A, the fulcrum is close to her and her force has a small lever arm x_g; in figure 11.2B, the fulcrum is close to the rock and the girl has a large lever arm. Suppose, in figure 11.2A, the fulcrum is only one foot from the girl's force, the rock will be eight feet from the fulcrum.

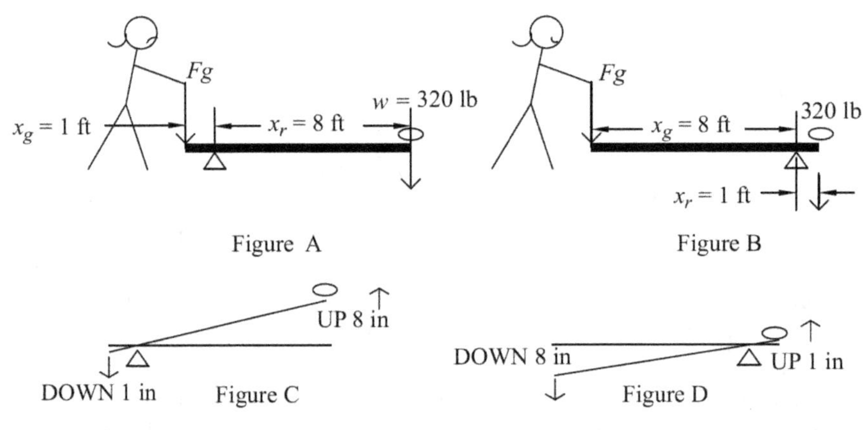

Figure 11.2. (A,B,C,D) Lifting a rock.

In figure 11.2A, the rock creates a clockwise torque about the fulcrum:

$$\tau_{cw} = 8 \text{ ft} \times 320 \text{ lb}$$

The girl must exert a downward force F_g to create a counterclockwise torque:

$$\tau_{ccw} = 1 \text{ ft} \times F_g$$

If the rock is lifted slowly with minimum acceleration, the torques must be equal:

$$\tau_{cw} = \tau_{ccw}$$

Therefore:

$$8 \text{ ft} \times 320 \text{ lb} = 1 \text{ ft} \times F_g$$

Solving for F_g, the girl must exert a downward force of 2560 lb! An unlikely scenario.

Far better is the arrangement shown in figure 11.2B with the fulcrum one foot from the stone: the torque equation now looks like:

$$\begin{aligned} 1 \text{ ft} \times 320 \text{ lb} &= 8 \text{ ft} \times F_g \\ F_g &= \frac{1 \text{ ft}}{8 \text{ ft}} \times 320 \text{ lb} = 40 \text{ lb!} \end{aligned}$$

The girl need only exert a force equal to one-eight of the rock's weight.

This example introduces a concept that helps us think and analyze forces and torques. This concept is the *mechanical advantage*—defined as the ratio of the girl's lever arm to the stone's lever arm or as the ratio of the weight of the rock to that of the girl's force:

$$\text{mechanical advantage} = \frac{\text{lever arm of girl's force}}{\text{lever arm of stone's weight}} = \frac{\text{weight of rock}}{\text{girl's force}} \quad (11.2)$$

In figure 11.2A:

$$\text{mechanical advantage} = \frac{1 \text{ ft}}{8 \text{ ft}} = 0.125$$

In figure 11.2B:

$$\text{mechanical advantage} = \frac{8 \text{ ft}}{1 \text{ ft}} = 8$$

Is there any advantage in trying to lift a stone with the small mechanical advantage such as figure 11.2A. Yes, there is a benefit; positioning the fulcrum close to the girl affords her a gain in the speed at which the rock is lifted. If she pushes the left edge of the board downward a vertical distance of one inch, the rock will be lifted upward a distance of eight inches in the same time interval (figure 11.2C). *A low mechanical advantage for force results in a large advantage for speed!* Figure 11.2B, with the high mechanical advantage for force, has a low advantage for speed (figure 11.2D).

It would take a super girl to lift the rock as in figure 11.2A; however, what if the rock only weighed a few pounds? In this case, trading the advantage for force is worth the gain in speed. We will shortly see that, in fact, most gearing on a bike operates at a low mechanical advantage for force, the trade being made for the advantage in speed.

11.4 Energy aspects of a high mechanical advantage

In the previous section, the girl employing a lever with a high mechanical advantage is able to lift a 320 lb stone by exerting a force of only 40 lb. There is an eight-fold gain in the force she is able to apply to the rock. Is she getting something for nothing? Is there some type of violation of the principle of conservation of energy? What is the magic here? There is no energy conservation violation and, sadly, no magic. Recall from chapter 8 that the work done is force multiplied by the distance moved in the direction of the force. If the girl exerts a downward force of 40 lb and moves the left end of the lever a distance of 8 ft, she performs an amount of work W equal to:

$$W = Fd = 40 \text{ lb} \times 8 \text{ ft} = 320 \text{ ft-lb}$$

Performance of work causes the potential energy of the stone to increase. The stone is raised upward a distance of one foot; thus, the 320 lb stone gains a potential energy equal to the weight w times the increase in height.

$$PE = w \times \Delta h = 320 \text{ lb} \times 1 \text{ ft} = 320 \text{ ft-lb}$$

The work done by the girl equals the gain in the stone's energy; there is no violation of energy conservation. The use of the lever has traded force for distance; the lever arrangement has also traded force for speed.

11.5 Multiple lever system

The concepts of the basic lever system can be expanded into systems of multiple levers. Maybe, the girl wishes to lift the rock in section 11.3 with a force less than 40 lb. While she could increase her mechanical advantage by using a longer bar or repositioning the fulcrum, there is another option: the use of multiple levers. Consider the combination of two levers (and two fulcrums) arranged to push on one another. The arrangement is illustrated in figure 11.3A.

To lift the rock at the right hand end of the second lever, the girl must lift upward on the first lever. This action causes a downward force on the second lever. The second lever provides further 'gain' in the girl's force. If both levers are configured with the mechanical advantage of eight, the 320 lb rock can be lifted with a force of:

$$F_g = \frac{1 \text{ ft}}{8 \text{ ft}} \times \frac{1 \text{ ft}}{8 \text{ ft}} \times 320 \text{ lb} = 5 \text{ lb!}$$

The multiple lever arrangement might seem awkward and a reasonable question is why not use a single board 18 ft long? Two boards can be arranged in a compact

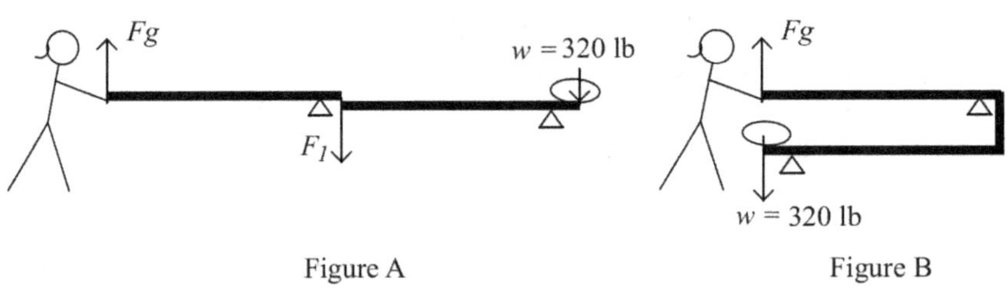

Figure 11.3. (A,B) Multiple levers.

manner (figure 11.3B). We will shortly see that the bike acts as a series of multiple, connected levers.

11.6 Early direct drive bicycles

The propulsion of a bicycle is a nice example of torque physics. Before delving into the complexities of chains and gears, consider a front wheel driven machine such as an early boneshaker (figures 2.3 and 2.4). These two-wheelers used a pedal directly connected to the front wheel (see figure 11.4).

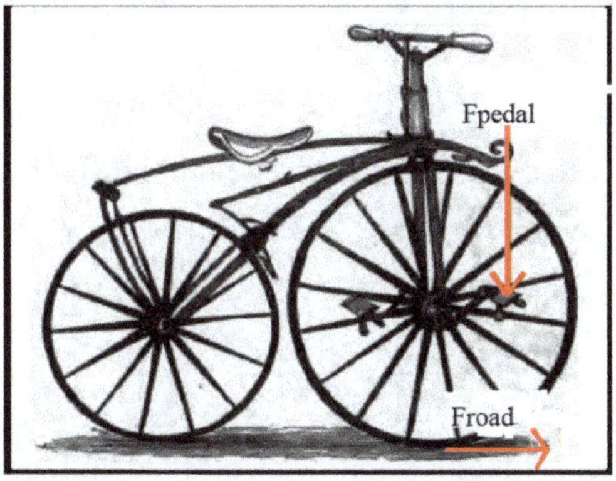

Figure 11.4. Forces on Boneshaker Velox 52.jpg [1].

The drive mechanism could not be simpler: the rider's force F_{pedal} pressing on the pedal creates a clockwise torque about the front axle with the crank length as a lever arm l_c.

$$\tau_{\text{cw}} = \text{force of rider pressing on pedal} \times \text{length of crank} = F_{\text{pedal}} \times l_{\text{c}}$$

When the wheel tries to spin clockwise, the ground exerts a reaction force F_{road} on the wheel. The reaction force produces a counterclockwise torque with the lever arm equal to the wheel radius r_w

$$\tau_{\text{ccw}} = \text{force of road} \times \text{radius of wheel} = F_{\text{road}} \times r_{\text{w}}$$

If the angular acceleration of the wheel is small, the clockwise and counterclockwise torques should be almost balanced:

$$\tau_{\text{cw}} = \tau_{\text{ccw}}$$
$$F_{\text{pedal}} \times l_{\text{c}} = F_{\text{road}} \times r_w$$

solving for the ground force:

$$F_{\text{road}} = \frac{F_{\text{pedal}} \times l_{\text{c}}}{r_{\text{w}}} \tag{11.3}$$

For a boneshaker with wheel radius of 15 in and a crank length of 7.5 in, a pedaling force of 50 lb from the rider results in a road force of:

$$F_{\text{road}} = \frac{50 \text{ lb} \times 7.5 \text{ in}}{15 \text{ inches}} = 25 \text{ lb}$$

The machine operates at a force mechanical advantage of:

$$\text{mechanical advantage} = \frac{25 \text{ lb}}{50 \text{ lb}} = 0.500$$

One-half of the rider's force on the pedal appears at the ground to propel the machine. The poor mechanical advantage for force results in a good advantage for speed. When the pedal makes one revolution, the foot travels a distance equal to the circumference of the pedal's circular path with a radius equal to the crank length:

$$C_{\text{pedal}} = 2\,\pi\,l_c = 2 \times 3.14 \times 7.5 \text{ in} = 47.1 \text{ in}$$

In this one revolution, the bike travels the circumference of the front wheel:

$$C_{\text{wheel}} = 2\,\pi\,r_w = 2 \times 3.14 \times 15 \text{ in} = 94.2 \text{ in}$$

The ratio of bicycle travel to pedal travel is:

$$\frac{C_{\text{wheel}}}{C_{\text{pedal}}} = \frac{94.2 \text{ in}}{47.1 \text{ in}} = 2.0$$

The reward for the rider is that his poor mechanical advantage of 0.5 for force results in a doubling of distance travel and speed.

11.7 High-wheelers

What are those Victorian high-wheel bikes all about? Why don't we use them now?

In a direct drive bicycle, such as the boneshaker or high-wheel (figures 2.3 and 2.6), the rotary crank was directly attached to the bike's driving wheel. Each turn of the crank caused the tire to make one revolution; the distance traveled was determined by the wheel's circumference. To increase the distance and speed per crank revolution, it was necessary to make the wheel larger. The 'High-wheelers' were described in terms of the wheel diameter, typically in the range of 50–60 inches. For instance, with a 52-inch front wheel, one turn of the crank moves the bicycle a distance:

$$\text{Circumference} = \pi \times \text{diameter} = 3.14 \times 52 \text{ inch} = 163 \text{ inch} = 13.6 \text{ ft}$$

A pedaling cadence of 60 revolutions per minute results in a speed of 13.6 ft/s about 9.3 miles per hour.

A 60-inch front wheel bicycle travels nearly 16 ft with each rotation of the pedal. At some point, the size of the front wheel is limited by the inseam distance of the rider. The reader is encouraged to perform a force analysis of the high-wheeler similar to that done for the boneshaker in section 11.6. The cranks in the high-wheels are about 6.5 inch (165 mm).

This focus on the diameter of the front wheel leads to a parameter called 'gear inches' which, for a direct drive machine, is the diameter of the driving wheel. The 60-inch diameter high-wheel is said to operate at '60 gear inches'. While this nomenclature might seem outdated, it was later used in bicycles that employed levers and gears to increase the wheel rotation and travel obtained from a single turn of the crank. Even today, units of 'gear inches' describe the gearing arrangement of modern machines.

There are several significant disadvantages to these direct drive high-wheelers. The rider is perched atop a 50 to 60 inch wheel leading to the possibility of a disastrous 'header' over the handlebars. The bicycle is difficult to propel uphill; the rider is essentially stuck in one gear—a high gear. Also, since the size of the big wheel is determined by the length of the rider's leg, a shorter rider is at a disadvantage to a taller rider. A further complication of the direct drive machine is that the pedals constantly turn; the rider must take his feet off the pedals and place them on a footrest to get relief while coasting. These disadvantages were addressed with a series of innovations now common on modern bicycles.

11.8 The safety bicycle

After the development of the high-wheeled bicycle, ingenious thinkers began to devise lever and gearing systems that would allow bikes to have smaller wheels that, nevertheless, matched the speed advantages of the big wheeler. Machines such as the Xtraordinary and the Facile were produced. These bicycles employed levers attached to the cranks such that one rotation of the pedal produced more than a single turn of the front wheel. Another bike, the Kangaroo, employed a chain connected set of gears, the same approach as in a modern bicycle. The Kangaroo had a wheel diameter of just 36 inches but one turn of the pedal produced a front wheel travel of 60 inches. While these bicycles offered increased gear inches without the need for immense drive wheels, they were soon rendered obsolete by the aptly named 'Safety' bicycle. This machine, essentially similar to a modern two-wheeler, employed relatively small wheels with the rider in a low position midway between the front and rear wheel. The pedals and cranks were attached to a front gear connected via chain to a rear gear and wheel. The most successful of these early safeties were the Rovers offered by Messrs. Starley and Sutton (figure 2.8).

The need for a large driving wheel was eliminated by use of gear ratios; the rotating crank is attached to a large front gear, the *chainring*. This chainring is connected with a chain drive to a small rear gear, the *cog*. The ratio of the chainring teeth to the number of cog teeth determines the rotation of the back tire for each turn of the crank. For instance, if the front ring had 48 teeth and the rear gear had 24 teeth, then one turn of the front crank would cause two turns of the rear gear. A bicycle with a 26-inch rear tire would travel a distance equivalent to a 52-inch wheel. Modern racing cycles operate in upper gears of 140 gear inches or more.

11.9 Force transmission in a geared bicycle

To understand the bicycle gearing, it is helpful to ignore the chain and imagine that the front chainring and rear cog are directly meshed. While the simplicity of such an arrangement might appeal to those who dislike cleaning and lubricating dirty chains, this scheme would only be useful on a very short bike that traveled backwards as the rider pedaled forwards! Figure 11.5 illustrates the direct meshing of the large chainring, radius R_{cr}, with the small cog, radius R_{co}.

The gears revolve about their centers and the gears' radii are the lever arms of the torques produced by forces between the gears. Since the teeth have to mesh with the chain, and in effect with each other, the front chainring and rear cog must have the same tooth spacing. Thus, the number of teeth is proportional to the diameter of the gear and, in effect, the actual gear diameters are not important—only the ratio of the number of teeth in the front chainring to the number of teeth in the rear cog. The role of the chain is to transmit via tension the force from the front chainring to the rear cog. The transmission of the pedaling force to the system's acceleration force may be considered as a series of steps (figure 11.6A):

1. The rider presses down on pedal thereby producing a torque to rotate the crank; the length of the crank is the lever arm.
2. The crank is attached to one or several front chainrings.
3. The front chainring rotates due to crank's torque—as it does, it develops a tension F_1 in the upper portion of the chain and the chain is pulled forward.
4. The forward motion of chain transmits the tension to the rear cog making it rotate.

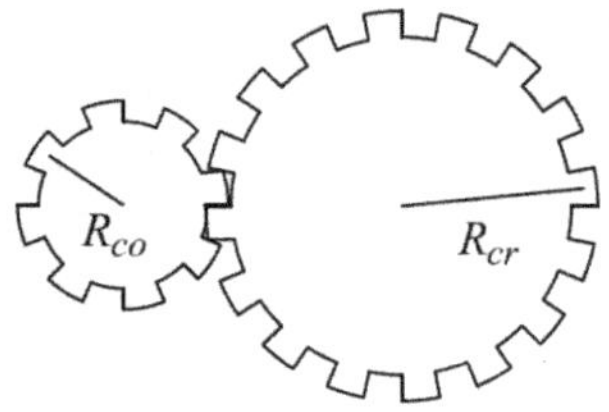

Figure 11.5. Gears.

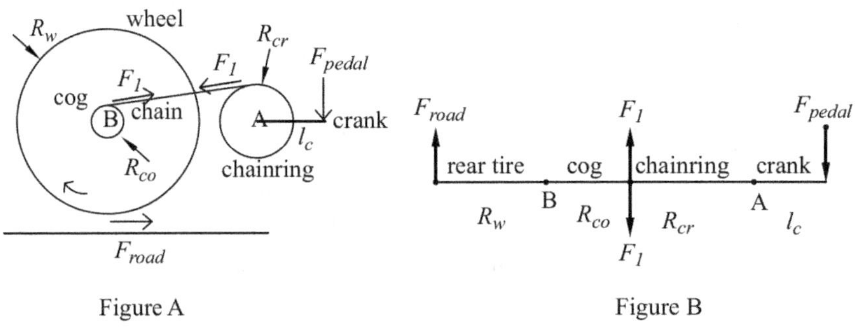

Figure 11.6. (A,B) Drivetrain.

5. The rear cog's rotation is transferred to the hub.
6. The hub pulls on the spokes.
7. The spokes pull on the wheel rim.
8. The tire, snugly attached to the wheel, tries to rotate and thus puts a rearward, horizontal friction force on the road.
9. The road exerts an equal and opposite horizontal friction force on the tire. This forward force propels the machine.

The bicycle's drive system is visualized as a multilever system (figure 11.6B).

As the rider presses down with a force on the pedal F_{pedal}, a clockwise torque is created on the front chainring; the lever arm of the force on the pedal is the length of the crank l_c. As the chainring tries to rotate about point A, it produces a tension F_1 in the chain. The tension's lever arm is the radius of the chainring R_{cr}. By looking at the torques on the front chainring, we write:

$$F_{\text{pedal}} \times l_c = F_1 \times R_{cr} \tag{11.4}$$

solving for F_1:

$$F_1 = \frac{F_{\text{pedal}} \times l_c}{R_{cr}} \tag{11.5}$$

The chain's tension is transmitted to the rear cog and causes a clockwise torque on the cog. At the cog, the lever arm of the chain tension is the radius of the cog R_{co}. Since the cog is attached to the rear wheel, the rear wheel tries to spin clockwise about point B and, therefore, exerts a rearward force on the ground. From Newton's Third Law, the ground exerts an equal and opposite force on the wheel F_{road}. This reaction force, directed toward the front of the bike, is the force of acceleration.

This ground force produces a counterclockwise torque. The lever arm for this force is the radius of the tire. For the torques on the rear wheel:

$$F_1 \times R_{co} = F_{\text{road}} \times R_w \tag{11.6}$$

Solving for F_{road}

$$F_{\text{road}} = \frac{F_1 \times R_{co}}{R_w} \tag{11.7}$$

Using equation (11.5) to substitute for F_1 and rearranging the fractions a bit:

$$F_{\text{road}} = F_{\text{pedal}} \times \frac{l_c}{R_w} \times \frac{R_{co}}{R_{cr}} \tag{11.8}$$

The last fraction is the ratio of the cog and chainring radii. Since the chainring and cog gears have the same tooth spacing, we write this fraction as the ratio of the number of cog teeth n_{co} and chainring teeth n_{cr}.

$$\frac{R_{co}}{R_{cr}} = \frac{n_{co}}{n_{cr}} \tag{11.9}$$

Thus:

$$F_{\text{road}} = F_{\text{pedal}} \times \frac{l_c}{R_w} \times \frac{n_{co}}{n_{cr}} \tag{11.10}$$

11.10 Multispeed gearing—force analysis

The result of the last section equation (11.10) is now applied to road and mountain machines equipped with multiple chainrings and cogs.

Road bikes

A road bicycle is usually outfitted with two chainrings and eight to ten cogs. Typical gear tooth counts might be:

Front chain rings	53 and 39 teeth
Rear cogs	10 to 26 teeth

The highest gear is obtained by using the largest chain ring in the front and the smallest cog in the rear. The lowest gear occurs with the smallest chainring and the largest cog. Some machines are equipped with a third small chainring of about 30 teeth—often referred to as the 'Granny gear'; this small chainring provides the rider with an additional range of low gears, especially useful for hills.

On adult bicycles, the length of crank ranges from 170 mm to 175 mm; we will use an intermediate crank length of 172.5 mm (6.79 inches). Wheel size varies slightly depending on type tire and inflation pressure; a wheel radius of 13.5 inch is common.

Pedaling in 'high gear' occurs with largest chainring of 53 teeth and smallest rear cog with 10 teeth. Substiting the various parameters into equation (11.10):

$$F_{\text{road}} = F_{\text{pedal}} \times \frac{l_c}{R_w} \times \frac{n_{co}}{n_{cr}} = F_{\text{pedal}} \times \frac{6.79 \text{ in}}{13.5 \text{ in}} \times \frac{10 \text{ teeth}}{53 \text{ teeth}} = 0.0949 \times F_{\text{pedal}} \tag{11.11}$$

A rider's perpendicular force of 50 lb on the pedal produces a forward force from the road of:

$$F_{\text{road}} = 0.0949 \times 50 \text{ lb} = 4.75 \text{ lb}$$

An alternative way to consider the relationship between the pedal and road force is to rearrange the equation (11.11) result and obtain:

$$F_{\text{pedal}} = 10.5 \times F_{\text{road}} \tag{11.12}$$

This gearing arrangement results in the bike operating at a low mechanical advantage for force; only about 10% of the rider's effort is delivered as force to ground. The equal and opposite reaction from the ground is the force of acceleration. This is expected since the *purpose of high gear is to allow for maximum speed.*

The lowest gear for this road bike occurs with a chainring of 39 teeth and a cog of 26 teeth. The relation between pedal and road force is then:

$$F_{\text{road}} = F_{\text{pedal}} \times \frac{l_c}{R_w} \times \frac{n_{co}}{n_{cr}} = F_{\text{pedal}} \times \frac{6.79 \text{ in}}{13.5 \text{ in}} \times \frac{26 \text{ teeth}}{39 \text{ teeth}} = 0.335 \times F_{\text{pedal}} \quad (11.13)$$

upon rearranging:

$$F_{\text{pedal}} = 2.99 \times F_{\text{road}} \quad (11.14)$$

The question might arise as to the efficiency of the bicycle drive train—how much of the applied force is 'lost' in friction and other inefficiencies between the gears and the chain? Frank Berto and Chester Kyle performed careful measurements in a laboratory environment and concluded that properly maintained drive trains operate at efficiencies near 95% [2]. The losses are sufficiently small and are ignored in our calculations.

In many of the above equations, there is a term that is the ratio of the number of cog teeth to the chainring teeth; it is useful to write the inverse and call it the gear ratio:

$$\text{gear ratio} = \frac{n_{cr}}{n_{co}} \quad (11.15)$$

When the bike is in high gear, the gear ratio is:

$$\text{gear ratio} = \frac{53}{10} = 5.3$$

Making a substitution of equation (11.15) into equation (11.10):

$$F_{\text{road}} = F_{\text{pedal}} \times \frac{l_c}{R_w} \times \frac{1}{\text{gear ratio}} \quad (11.16)$$

Gear inches

Although the term *gear inches*, the diameter of the driving wheel in the Victorian ordinary, was best suited to comparing the big wheelers, it is still used by modern riders as a parameter describing their gear arrangement. In the notation of this section, this parameter is:

$$\text{gear inches} = 2 \times R_w \times \text{gear ratio} = 2 \times R_w \times \frac{n_{cr}}{n_{co}} \quad (11.17)$$

In high gear, our road bike operates in gearing of:

$$\text{gear inches} = 2 \times R_w \times \frac{n_{cr}}{n_{co}} = 2 \times 13.5 \text{ inch} \times \frac{53}{10} = 143 \text{ inch}$$

The road bike in high gear would have speed equivalent of a high-wheeler of diameter 143 inches. The high-wheel rider would need an inseam of 72 inches!

Development

While many cyclists evaluate their gearing in units of gear inches, another related and more informative parameter is the *development* of the machine. The development is the distance traveled with each revolution of the pedal—the product of the rear wheel circumference and the gear ratio:

$$\text{development} = 2\pi \times R_w \times \text{gear ratio} = 2\pi \times R_w \times \frac{n_{cr}}{n_{co}}$$

For our road bike in high gear:

$$\text{development} = 2\pi \times R_w \times \frac{n_{cr}}{n_{co}} = 2\pi \times 13.5 \text{ inch} \times \frac{53}{10} = 450 \text{ inch}$$

Nearly 38 ft in a single turn of the pedals—the magic of the bicycle!

A useful comparison can be made to the distance traveled by the pedal C_p during one revolution; it is the circumference of the crank:

$$C_p = 2\pi \times l_c = 2\pi \times 6.79 \text{ inch} = 42.7 \text{ inch} = 3.56 \text{ ft}$$

The ratio of distance traveled by the bike to that of the pedal:

$$\frac{450 \text{ inch}}{42.7 \text{ inch}} = 10.5$$

is instructive when compared to equation (11.12). In return for the rider having to exert a force on the pedal 10.5 times as large as the force on the road, she travels 10.5 times further. Once again, we see the principle of the simple lever (section 11.3).

Mountain bicycles

To apply the above concepts to mountain bicycles, a few changes must be made. Since mountain bikes have higher bottom brackets, slightly longer cranks are used when compared to road bikes; common is 175mm (6.89 inches). Mountain bikes also come with different wheel/tire diameters and wide tires; a 26-inch diameter is common, although some mountain bicycles now come with 29-inch diameter wheels. The most important difference in the mountain bike is in the gearing; the front gears are triple chainrings of smaller size than found on a road bike. For instance: front chainrings = 44/32/22 teeth and rear cogs = 11 to 34 teeth. Using these numbers we apply equation (11.10):

In high gear with a 44 tooth chainring and an 11 tooth cog:

$$F_{road} = F_{pedal} \times \frac{l_c}{R_w} \times \frac{n_{co}}{n_{cr}} = F_{pedal} \times \frac{6.89 \text{ in}}{13 \text{ in}} \times \frac{11}{44} = 0.133 \times F_{pedal} \qquad (11.18)$$

or

$$F_{pedal} = 7.52 \times F_{road} \qquad (11.19)$$

Table 11.1. Summarizes the various ratios of road force to pedaling force for both the road and mountain bicycles.

Gear	Road bicycle: $\frac{\text{chainring}}{\text{cog}}$		Mountain bicycle: $\frac{\text{chainring}}{\text{cog}}$	
High	53:10 = 5.3	$F_{\text{pedal}} = 10.5 \times F_{\text{road}}$ $F_{\text{road}} = 0.0949 \times F_{\text{pedal}}$	44:11 = 4.0	$F_{\text{pedal}} = 7.52 \times F_{\text{road}}$ $F_{\text{road}} = 0.133 \times F_{\text{pedal}}$
Low	39:26 =1.5	$F_{\text{pedal}} = 2.99 \times F_{\text{road}}$ $F_{\text{road}} = 0.335 \times F_{\text{pedal}}$	22:34 = 0.647	$F_{\text{pedal}} = 1.22 \times F_{\text{road}}$ $F_{\text{road}} = 0.819 \times F_{\text{pedal}}$

In low gear with a 22 tooth chainring and a 34 tooth cog:

$$F_{\text{road}} = F_{\text{pedal}} \times \frac{l_c}{R_w} \times \frac{n_{co}}{n_{cr}} = F_{\text{pedal}} \times \frac{6.89 \text{ in}}{13 \text{ in}} \times \frac{34}{22} = 0.819 \times F_{\text{pedal}} \tag{11.20}$$

or

$$F_{\text{pedal}} = 1.22 \times F_{\text{road}} \tag{11.21}$$

Table 11.1 summarizes the various ratios of road force to pedaling force for both the road and mountain bicycles.

The lowest gear in the mountain bike causes 81.9% of pedaling force to be applied to the road, whereas the lowest gear in the road bike results in only 33.5% of pedaling force to be applied to the road. At all gear ratios, the acceleration force from the ground is but a fraction of the rider's pedaling effort. Based on our understanding of the lever concepts from section 11.3, the low mechanical advantage of the pedaling force results in a gain in the bicycle's speed.

Are all of those gears on a bike really necessary?

In a bicycle with variable gears, two rings in the front and eight cogs in the rear would theoretically make the bike a *16-speed*. Three chain rings and ten rear cogs would afford the possibility of *30-speeds*. There are two caveats to the described terminology. No matter what the gearing, ultimately, the speed of the bicycle is determined by the rotary speed of the crank. Moreover, there is redundancy in the front to back tooth ratios, i.e. 48/24 would be the same gear as 24/12; other gear ratios might be very close. There are also certain gear combinations that are mechanically unwise, for instance, those that require a chain crossing from front to back at an extreme angle. That *30 speed* bike might have only 20 useful gear combinations! So, are all those gears really necessary? Ask the bike manufacturer's marketing department.

11.11 Gearing and pedaling cadence

By knowing the wheel radius R_w along with the gear ratio $\frac{n_{cr}}{n_{co}}$, it is straightforward to relate the speed of the bike v to the pedaling cadence. It is best to express all distances in feet. The speed of the cycle is the product of the pedaling cadence in revolutions per second ($\text{cadence}_{\text{rps}}$), the gear ratio, and the distance traveled during one revolution of the wheel (C_w, circumference of the wheel):

$$v = \text{cadence}_{\text{rps}} \times \frac{n_{\text{cr}}}{n_{\text{co}}} \times C_w \tag{11.22}$$

Using wheel radius R_w to obtain circumference $C_w = 2\pi \times R_w$ and by dividing by 60 to use a cadence in revolutions per minute ($\text{cadence}_{\text{rpm}}$), we obtain:

$$v = \frac{\text{cadence}_{\text{rpm}}}{60} \times \frac{n_{\text{cr}}}{n_{\text{co}}} \times 2\pi R_w \tag{11.23}$$

$$\text{cadence}_{\text{rpm}} = \frac{60 \times v}{(2\pi \times R_w) \times \dfrac{n_{\text{cr}}}{n_{\text{co}}}} \tag{11.24}$$

It is important to write both the bike speed and wheel radius in feet. A road bike traveling 20 mph (29.3 ft/s) with a wheel radius 13.5 inch (1.125 ft) in a high gear ratio of 5.3 (53/10) will require a pedaling cadence of:

$$\text{cadence}_{\text{in rpm}} = \frac{60 \times 29.3 \text{ ft/s}}{(2\pi \times 1.125 \text{ ft}) \times 5.3} = 47 \text{ rpm}$$

Traveling 50% faster will demand a cadence of $1.5 \times 47 \text{ rpm} = 71 \text{ rpm}$. Many riders will prefer even higher cadences and lower gear ratios.

11.12 Gearing and pedaling force

We now explore the relationship between the road and pedaling forces required to overcome factors such as resistance and slope forces.

Riding on level road at constant speed

Let us ponder further the cycling situation described at the end of the previous section—pedaling a road bike in high gear at a constant speed of 20 mph. To get an appreciation of the effort required of the rider, we use the results of table 8.3 and see that, at a speed of 20 mph, the total resistive force on our standard rider is 6.51 lb. A forward force from the road of 6.51 lb must counter this backward force. If riding in high gear, the relation of road force to pedaling force (Table 11.1) may be used:

$$F_{\text{pedal}} = 10.5 \times F_{\text{road}} = 10.5 \times 6.51 \text{ lb} = 68.4 \text{ lb}$$

A major effort by our standard rider of 161 lb! Actually, the situation is even worse; this calculated force is the *average force*. During the course of a pedal rotation, the relationship of the leg to the pedal causes the instantaneous force to vary. In the pedal rotation cycle, there are positions that require rider force higher than the average of 67.1 lb.

The speed obtained in high gear is at the expense of a large pedaling force; the use of a lower gear reduces pedaling force. If riding in low gear, the relation between road force to pedaling force (table 11.1) is:

$$F_{\text{pedal}} = 2.99 \times F_{\text{road}} = 2.99 \times 6.51\ \text{lb} = 19.5\ \text{lb}\ !$$

The downside of this low gear is that the rider has to maintain an unrealistic cadence to move the bike at 20 mph.

Pedaling power

It is instructive to consider the rider's effort from the perspective of power. Riding at 20 mph in high gear requires a pedaling force that averages 68.4 lb and, during one revolution of the crank, the pedal travels a distance equal to the crank's circumference of 42.6 in (3.55 ft). At a cadence of 47 rpm (0.783 rev/s), the speed of the pedal is:

$$\frac{3.55\ \text{ft}}{\text{rev}} \times \frac{0.783\ \text{rev}}{\text{s}} = 2.78\ \text{ft/s}$$

Recall from chapter 8, the power generated by a force moving a body at a given speed is equation (8.14):

$$\text{power} = \text{force} \times \text{speed} = Fv$$

Substituting numbers: $P = Fv = 68.4\ \text{lb} \times 2.78\ \text{ft/s} = 190\dfrac{\text{ft-lb}}{\text{s}}$

A comparison of the mechanical power output in table 8.3 shows that a 20 mph ride requires a mechanical power output of 191 $\dfrac{\text{ft-lb}}{\text{s}}$; within minor round off, the values match!

Using the pedaling force and the pedaling cadence to obtain mechanical power output provides an 'instantaneous' reading to the cyclist.

Climbing hills

In previous discussions, we established that high gears allow the bike to travel fast while maintaining a comfortable cadence. Let us now examine the role of gear choice for ascending hills.

In climbing even moderate hills, a substantial portion of the rider's effort goes toward overcoming the downward pull of gravity, the slope force (section 6.4). Table 6.1 illustrates that our standard rider ascending a modest 5% grade encounters a slope force of 9.04 lb. If our rider attacks this hill at 20 mph, she must deal with a total force equal to the slope force F_{slope} and the resistive force F_{tot} of 6.51 lb (table 8.3):

$$F_{\text{total}} = F_{\text{slope}} + F_{\text{tot}} = 9.04\ \text{lb} + 6.51\ \text{lb} = 15.6\ \text{lb}$$

Riding in high gear would require an average pedaling force:

$$F_{\text{pedal}} = 10.5 \times 15.6\ \text{lb} = 164\ \text{lb}!$$

Not so easy for our standard 161 lb rider! She would have to push with all of her weight on one pedal and simultaneously pull up on handlebars to further increase the downward efforts of her body. Much more likely is that she shifts to a lower gear and also reduces her speed. Perhaps, she drops into the road bike's lowest gear ratio of 1.5 (table 11.1) and maintains her cadence, from equation (11.23):

$$v = \frac{n_{cr}}{n_{co}} \times \frac{2\pi \times R_w \times \text{cadence}_{rpm}}{60} = 1.5 \times \frac{2\pi \times 1.125 \text{ ft} \times 47}{60} = 8.3 \text{ ft/s} = 5.66 \text{ mph}$$

A major advantage of the speed reduction is that the force of air resistance is greatly lessened. Using a speed of 6 mph, we obtain from table 8.3 a resistive force of 1.50 lb. Of course, the slope force remains the same but the sum is significantly reduced:

$$F_{total} = F_{slope} + F_{tot} = 9.04 \text{ lb} + 1.50 \text{ lb} = 10.5 \text{ lb}$$

The average pedaling force in low gear is then $F_{pedal} = 2.99 \times 10.5 \text{ lb} = 31.4 \text{ lb}$.

What can be more pleasant—climbing the hill at 6 mph, an easy 47 rpm with a moderate pedaling force!

The above analysis of the road bike is easy to adapt to the mountain bike by adjusting for the mountain bike's lower gear ratios, wheel size, and longer crank. All things being equal—and they are not—mountain bikes are heavier and equipped with robust tires of greater rolling resistance; the mountain bikes should be able to climb a hill roughly *twice as steep* as the road bike. The road bike is superior for its top speed in high gear. Aside from the gearing, the road bike has smoother tires and better rider aerodynamics.Another speed advantage of the road bike is its lower mass and, therefore, quicker acceleration.

Acceleration

What is the ideal gear for getting a bike up to speed as quickly as possible?

Now that we have looked at the pedaling forces necessary to ride at high speeds of 20 mph and climb moderate hills, it is instructive to consider these pedaling forces in situations of acceleration. Suppose our standard rider tries to start a road bike in high gear by standing and applying all of her 161 lb body weight (mass of 5.62 slugs) to one pedal. The reaction force from the road provides the acceleration; from table 11.1 the road force in high gear is found to be:

$$F_{road} = 0.0949 \times F_{pedal} = 0.0949 \times 161 \text{ lb} = 15.3 \text{ lb}$$

Starting from rest, the initial force of air resistance is small; we also ignore the small resistance forces of the bearings and the rolling resistance. The acceleration is calculated from Newton's Second Law:

$$a = \frac{F}{m} = \frac{15.3 \text{ lb}}{5.62 \text{ slugs}} = 2.72 \text{ ft/s}^2 = 0.085 \text{ `g'}$$

A very small acceleration.

The acceleration produced by the same pedal force in low gear is also obtained from table 11.1:

$$F_{road} = 0.335 \times F_{pedal} = 0.335 \times 161 \text{ lb} = 53.9 \text{ lb}$$

and the acceleration:

$$a = \frac{F}{m} = \frac{53.9 \text{ lb}}{5.62 \text{ slugs}} = 9.59 \text{ ft/s}^2 = 0.30 \text{ ‘g’}$$

The low gear provides a much larger initial acceleration. As the bike picks up speed, the rider shifts into higher gears to maintain a reasonable cadence. The use of a low gear for starting out is well known to anyone who has operated a manual transmission car.

Suppose our rider wishes an even greater starting acceleration—she might pull up on the handlebars and, by Newton's Third Law, the handlebars would pull down on her. This downward force from the handlebars adds to the force she delivers to the pedals (section 5.5). If she is strong and capable of pulling up on each grip with a force of 40 lb, the total upward force on handlebars is 80 lb. The resultant force on the pedal is

$$F_{\text{pedal}} = w + F_{\text{handelbars}} = 161 \text{ lb} + 80 \text{ lb} = 241 \text{ lb}$$

In the lowest road bike gear:

$$F_{\text{road}} = 0.335\ F_{\text{pedal}} = 0.335 \times 241 \text{ lb} = 80.7 \text{ lb}$$

For the acceleration:

$$a = \frac{F_{\text{road}}}{m} = \frac{80.7 \text{ lb}}{5.62 \text{ slugs}} = 14.4 \text{ ft/s}^2$$

This is almost 0.5 ‘g’. Our rider is a ‘rocket woman’! However, caution is needed in accepting this number.

Consider that magnitude of the road force $F_{\text{road}} = 80.7$ lb and the source of this force is static friction between the rear tire and the road. From chapter 5, the expression for the maximum force of static friction:

$$f_{\text{max}} = \mu_{\text{static}} N$$

where μ_{static} is the coefficient of static friction and N is the normal force between the road and the rear tire.

For a seated rider, approximately 60% of the system weight is on the rear tire ($N = 0.6 \times 181$ lb $= 108.6$ lb). Standing will shift these weight proportions, but to simplify the analysis, let us keep our rider in a crouched position and use the 60/40 ratio. Using a coefficient of friction of 0.9 gives:

$$f_{\text{max}} = 0.9 \times 108.6 \text{ lb} = 97.7 \text{ lb}$$

These numbers illustrate that, with a slightly smaller coefficient of static friction or a greater upward pull on the handlebar, the rear tire is close to spinning. The same effect is observed when an automobile driver ‘gives it too much gas’ when trying to pull out of an icy parking spot. Another possible complication that develops when large forces are applied to the bike's back wheel is the ‘wheelie’, i.e., the lifting of the front tire. This wheelie effect is evaluated in a later section with a torque calculation.

It is worth noting that no amount of pulling (or pushing) on the handlebars is going to change the system weight of 181 lb and, therefore, the total vertical force exerted on the ground. The forces she exerts on the handlebars or transfers to a pedal by standing are *internal* forces in the system defined as 'bike + rider'. The only way that the rider could increase the downward force onto the ground would be to stand and then somehow crash onto the seat; try jumping on a scale to see this effect. Of course, the increase in weight is brief and is preceded by a reduction in weight when the rider initially lifted herself upward.

The considerations of this section show that maximum acceleration will not be obtained in the highest gear—but if the gear ratio is too low, the wheels might spin—so what works best? Probably a fairly low gear, but maybe not the absolute lowest.

11.13 Braking

Are both front and rear brakes equally effective at stopping the bike?

In a later section, we establish that during a hard stop a large fraction of the system weight is transferred to the front tire. This reduction of the normal force at the rear tire will proportionally reduce the rear tire's stopping effectiveness.

Which type of brake is best—rear coaster or hand lever?

We suspect that rear coaster brakes were introduced out of concern that children might not possess sufficient grip strength for handlebar lever brakes. While rear coaster brakes are not as common as they were 50 years ago, they are still found on contemporary bicycles for young (3–6 years old) children. Another benefit is that rear coaster brakes have good immunity to dirt and damage. They also have significant disadvantages including:

- The aforementioned reduced stopping effectiveness during a hard stop
- Young children get confused in making the transition from the forward pedaling motion to the rearward actions needed to stop with a coaster brake
- The rearward braking motion is ineffective when the cranks are near the vertical, another issue with kids
- Poor mechanical advantage due to small lever arm of hub
- Loss of braking if foot slips off pedal or the chain breaks
- Complex disassembly and repair
- Difficult inspection for wear or broken parts

The rear coaster brake action occurs inside the rear hub and cannot be observed; we will not examine the physics of the coaster's operation. The reader may be assured that there are torques involved in reducing the rotational speed of the wheels.

Why does it take the large muscles of our lower body to accelerate the bike yet we can stop with the forces developed by our relatively small hands?

The efforts of the hands operate at a high mechanical advantage for force. The forces of propulsion function at a low mechanical advantage for force; the trade-off being that the propulsion forces offer a gain in speed. We begin our analysis of bicycle braking by looking at the handlebar levers, the first component in a modern road or mountain bike braking system.

Brake levers

The brakes on a bicycle reveal another application of torques to cycling. When the cyclist decelerates the machine with hand brakes, there are several locations where torques are important:

- The hand levers—where forces exerted by the fingers, produce torques causing a rotation of the lever and the development of tension in brake cable wire.
- The caliper brakes—that rotate as a result of the torque produced by cable wire tension.
- The rims or disk brake rotors—when squeezed by pads cause a torque that slows the wheel's rotation.
- The contact between the tire and the road—where the road's force of friction produces a rearward pointing force. This horizontal force of the road on the bike is the cause of the machine's deceleration.

The grip forces in hand brakes have a large mechanical force advantage due to the brake lever's geometry. In both the vertical levers found in road bikes and the horizontal levers typical of mountain and hybrid bikes, we find substantial mechanical advantages.

Figure 11.7 illustrates a right side horizontal bike lever. Squeezing the brake lever with the hand produces a clockwise torque about the lever's point of rotation; the lever pulls on the cable and the cable produces an equal and opposite force on the lever. The tension in the cable results in a counterclockwise torque on the lever. (Although the hand's force is usually delivered by three or four fingers, we have represented the force as a single vector.) The finger's lever arm is 3.0 inch and the cable's lever arm is 1.0 inch. Although the maximum grip strengths of most adults is surprisingly high, with a range from about 60 to 110 lb [3]; we will use a modest hand grip force of only 20 lb. Once the brake pads contact the rim, the rider continuously holds the lever in position and there is a balance of torques about the point of rotation:

$$t_{\mathrm{cw}} = t_{\mathrm{ccw}}$$

$$3.0 \text{ inch} \times 20 \text{ lb} = 1.0 \text{ inch } \times T_{\mathrm{c}}$$

Solving for cable tension:

$$T_{\mathrm{c}} = 60 \text{ lb}$$

The hand force operates at mechanical advantage of 3X. While this is an impressive tension in the cable, the braking force is amplified further by the brake pads at the rim.

Rim brakes

There are several types of rim brakes—these include center pivot side pull calipers, dual pivot side pull calipers, center pull, and cantilevered 'V' brakes (linear pull). The various types differ in construction, ease of adjustment, reach of the brake

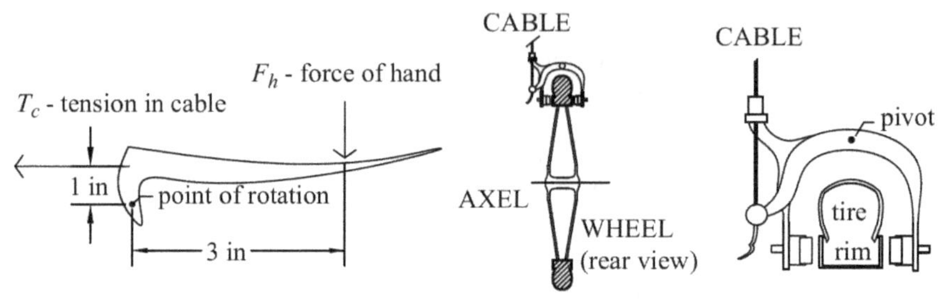

Figure 11.7. (A,B,C) Brake system.

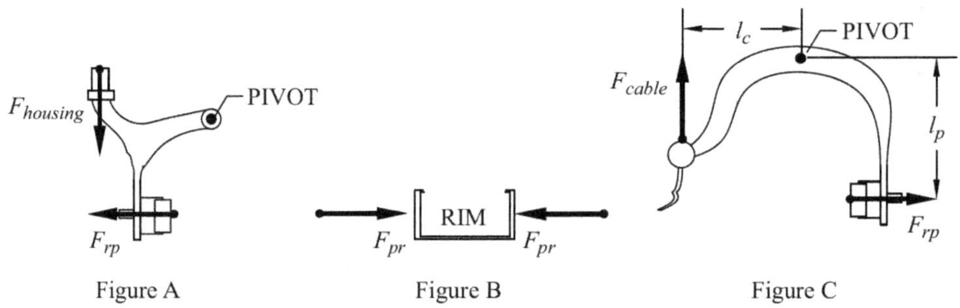

Figure 11.8. (A,B,C) Caliber brake exploded.

pads, and 'grabbing force'; they all offer an increased mechanical advantage to the cable's tension. Figures 11.7B and 11.7C illustrate the center pivot side pull caliber brake.

The brake consists of two asymmetrically shaped arms upon which are mounted the left and right hand brake pads. The tension in the brake cable results in a force being applied to the two caliper arms (figure 11.7C). As the cable pulls up on one arm, the cable housing pushes down on the other arm. The individual brake arms are shown in an exploded view in figures 11.8A and 11.8C. The force on the left arm F_{housing} and the force on the right arm F_{cable} will each be one-half of the tension in the brake cable. We have previously shown that a 20 lb squeeze on the brake lever will result in a cable tension of 60 lb; the cable tension can be split between the housing and cable resulting in a force of 30 lb on the respective brake arms.

As the cable develops tension, it pulls upward on the right side ('u' shaped) arm (fgure 11.8C); this arm rotates clockwise causing the pad to press against the rim with a force F_{pr}. The rim then exerts an equal and opposite force on the brake pad F_{rp}. A similar pair of action–reaction forces develop on the 'y' shaped left side caliber arm (figure 11.8A). Since the 'u' shaped brake arm is on the outside, it is easier to observe its operation and measure the lever arms of the forces. Writing the torques on the right pad brake arms (figure 11.8C)

$$t_{\text{cw}} = t_{\text{ccw}}$$
$$l_{\text{c}} \times F_{\text{cable}} = l_{\text{p}} \times F_{\text{rp}}$$

Solving for the force exerted on the rim by the brake pad:

$$F_{rp} = \frac{l_c \times F_{\text{cable}}}{l_p}$$

By Newton's Third Law, the force of the pad on the rim F_{pr} will be equal and opposite. The geometry of the left and right brake arms will determine the forces' lever arms. Using values of 3 inch for the cable tension's lever arm and 1.5 inch for the pad force's lever arm, a cable tension of 30 lb results in a pad force F_{pr}:

$$F_{\text{pr}} = \frac{l_c \times F_{\text{cable}}}{l_p} = \frac{3 \text{ inch} \times 30 \text{ lb}}{1.5 \text{ inch}} = 60 \text{ lb}$$

The rim is squeezed from both sides with this magnitude. This force is the 'normal' or perpendicular force in the frictional interaction with the rim. The frictional force is a vector tangent to the rim's circumference (figure 11.9). Since the wheel's radius is the force's lever arm, a large torque is possible at the rim.

The frictional contact between the brake pad and the rim is that of sliding friction. The coefficients of friction will vary depending on specific combinations of pad and rim materials. As long as the contact surfaces are clean and dry, the frictional coefficients will be high, typically 0.6 or greater.

Using this value, we write the frictional force from one pad as:

$$f = \mu_k \times F_{\text{pr}} = 0.6 \times 60 \text{ lb} = 36 \text{ lb}$$

With a pad on each side, the total frictional force on the rim is double this value. There is ample force available to stop tire rotation.

Since the rim is spinning, the force is sliding friction. Sliding friction always generates heat; on bikes with metallic rims, the heat is well dissipated; however, the heat can be a problem with carbon rims.

The forces of the brake pads on the rims are *internal forces* to the bike/rider system; they do not decelerate the system. Only net *external forces* can change an object's velocity. The brake pad forces create a torque that reduces the rotational

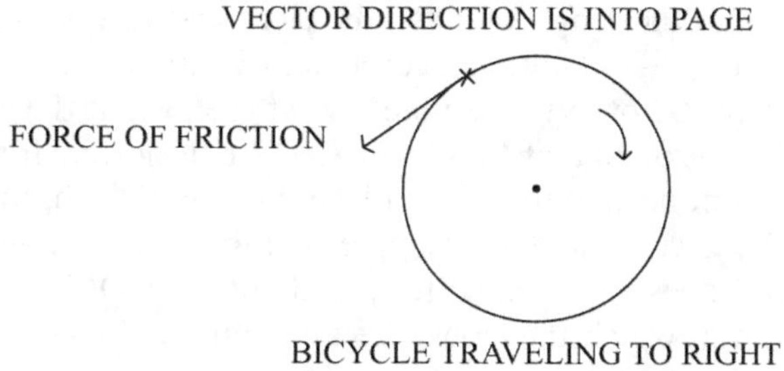

Figure 11.9. Brake forces on rim.

speed of the wheel. In section 5.9, we saw that reducing the rotation of the wheel causes the wheel to push on the ground with a forward pointing frictional force. The ground then pushes with a rearward friction on the wheel. *This backward push from the ground is the deceleration force.* The primary limitation in decelerating the bicycle is the frictional contact between the tire and the road and the necessity to avoid large decelerations that trigger headers. In the absence of skidding, both the acceleration and deceleration of the bicycle are due to static friction with no generation of heat between the wheels and the ground.

11.14 Wheelies

We now evaluate the 'wheelie'—the lifting of the front tire during a large forward acceleration. The wheelie is a consequence of large torques developed at the rear tire as a rider pedals hard in a low gear. For the discussion of the wheelie, and headers in a subsequent section, we use an analysis and geometry similar to that of Wilson [4].

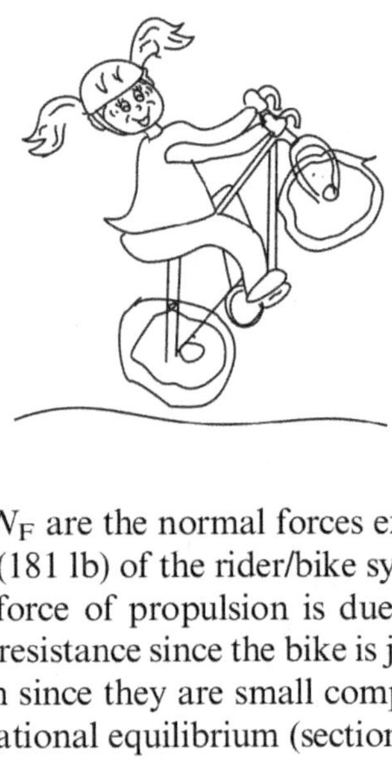

In (figure 11.10), N_R and N_F are the normal forces exerted by the road on the rear and front wheels; the weight (181 lb) of the rider/bike system is shown at the center of mass (COM). The forward force of propulsion is due to friction at the rear wheel F_{road}. There is no force of air resistance since the bike is just starting to move; the other resistive forces are not shown since they are small compared to the propulsion force. We apply the concepts of rotational equilibrium (section 11.3) and sum torques about the system center of mass. With the COM as the point of torque computation, the frictional force on the rear wheel produces a torque that is trying to rotate the bike counterclockwise about the center of mass; the clockwise torque is due to the vertical normal force on the rear. An additional counterclockwise torque is due to the normal force on the front wheel; however, we note that at the instant the wheelie begins, the front tire lifts from the ground and N_F vanishes. The rear normal force has a lever arm of 17 inch and the front normal force has a lever arm of 25 inch; the road force has a lever arm of 45 inch. Since the system weight of 181 lb passes through the COM, it has a zero lever arm and does not create a torque about the COM.

For the condition in which the clockwise and counterclockwise torques cancel:

$$\tau_{cw} = \tau_{ccw} \tag{11.25}$$

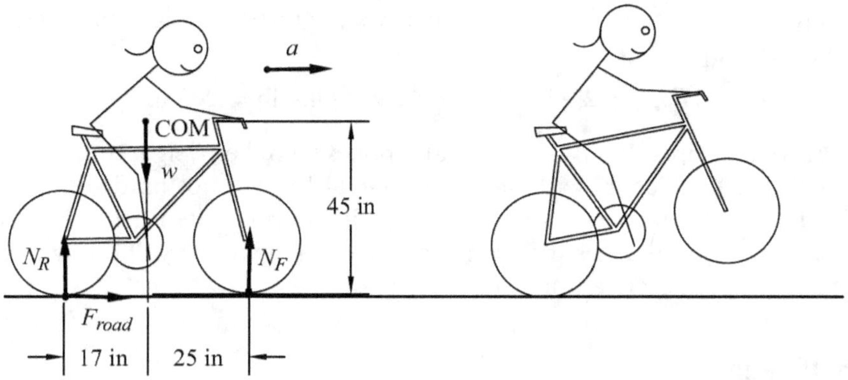

Figure 11.10. Wheelie.

thus:

$$N_R \times 17 \text{ inch} = (N_F \times 25 \text{ inch}) + (F_{\text{road}} \times 45 \text{ inch})$$

Consider the vertical forces on the system: since there is no vertical acceleration, the forces sum to zero:

$$N_R + N_F - 181 \text{ lb} = 0$$

The wheelie develops when the front wheel loses all contact with the ground and therefore:

$$N_F = 0$$

All of the system weight is now on the rear wheel:

$$N_R = 181 \text{ lb}$$

Putting these numbers into the torque expression (11.25):

$$181 \text{ lb} \times 17 \text{ inch} = (0 \times 25 \text{ lb}) + (F_{\text{road}} \times 45 \text{ inch})$$

and solving for the road's force of friction:

$$F_{\text{road}} = \frac{181 \text{ lb} \times 17 \text{ inch}}{45 \text{ inch}} = 68.4 \text{ lb}$$

The acceleration that induces the wheelie may be calculated:

$$a = \frac{F_{\text{road}}}{m} = \frac{68.4 \text{ lb}}{5.62 \text{ slugs}} = 12.2 \frac{\text{ft}}{\text{s}^2} = 0.379 \text{ ‘g’}$$

We can think of the ‘cause’ of the wheelie as excessive, forward pointing, frictional force on the bottom of the rear wheel.

Does such a magnitude of frictional force require a large coefficient of friction? It is easy to calculate:

$$\mu = \frac{f}{N_R} = \frac{68.4 \text{ lb}}{181 \text{ lb}} = 0.378$$

This is a relatively low coefficient; recall from table 5.1 that the coefficient of static friction between rubber and blacktop is about 0.9. To obtain the pedal force required to produce the wheelie in low gear, we use the expression (table 11.1) from earlier in this chapter:

$$F_{pedal} = 2.99 \times F_{road} = 2.99 \times 68.4 \text{ lb} = 205 \text{ lb}$$

If our 161 lb rider rises slightly off the seat, applies all of her weight to the pedal, and pulls up on the handlebars with an additional 44 lb, it is not hard to produce the wheelie! The reader may wish to verify that, if the rider sits further back on the bicycle and, thereby, reduces the lever arm of N_R, the wheelie will occur at a lower pedaling force and acceleration.

11.15 Headers

The header is the opposite of the wheelie in both cause and action. The wheelie is triggered during a rapid start by an excess of forward directed, frictional force on the rear wheel. The header is a consequence of excessive, rear directed, frictional braking force on the front wheel. Hard braking with the front wheel creates the most dangerous condition. While there is no joy on a bike when going headfirst over the handlebars, the event involves interesting physics. Perhaps, if the reader is convalescing with a broken collarbone, she can study this section to better understand and avoid the header. The basic physical event for somersaulting the rider over the handlebars occurs when the overall system rotates in the forward direction. This rotational scenario, the classical *header*, develops without colliding with any obstacle.

The intuitive explanation of the header is that, with increasing decelerations, the rear tire carries less and less of the system weight. At some point, this force on the rear tire is zero; the back wheel has lifted off the ground. We now have the rider in the dangerous condition for a header over the handlebars.

Figure 11.11A illustrates the forces on a rider who is applying the brakes. There are normal (vertical) forces and ground friction at the bottom of each tire. All torques are computed about the COM. The system weight has a zero lever arm. Another force, air resistance, acts near the rider's mid-section; its lever arm is small and the air resistance torque is ignored. The indicated distances are the lever arms of the forces as measured from the COM.

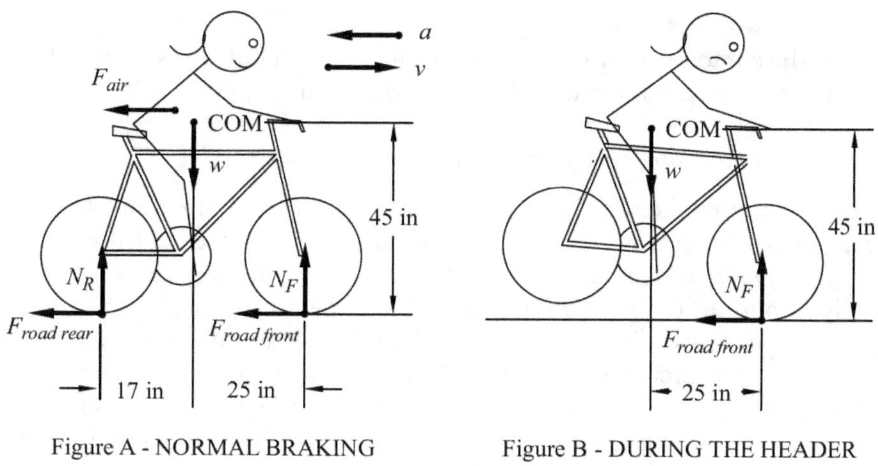

Figure 11.11. (A, B) Headers.

The braking forces exert a clockwise torque about the COM; this torque results in the rear wheel lifting and weight being transferred to the front tire (figure 11.11B). The front normal force must equal the system weight and all stopping force takes place at the front tire.

We have depicted an upright rider; in reality, the rider might change lever arms by slipping further back on the seat and also bend and lower her center of mass.

The torque equation at the onset of the header:

$$\tau_{\text{cw}} = \tau_{\text{ccw}}$$

$$F_{\text{road front}} \times 45 \text{ inch} = N_{\text{F}} \times 25 \text{ inch}$$

Since the front wheel supports entire system weight: $N_{\text{F}} = 181$ lb.

We obtain for the frictional force triggering the header

$$F_{\text{road front}} = \frac{181 \text{ lb} \times 25 \text{ inch}}{45 \text{ inch}} = 101 \text{ lb}$$

The deceleration at onset of header is:

$$a = \frac{F}{m} = \frac{101 \text{ lb}}{5.62 \text{ slugs}} = -18.0\frac{\text{ft}}{\text{s}^2} = -0.559\text{ ‘g’}$$

The ‘cause’ of the header is too large of a rearward frictional force on the bottom of the front wheel.

What coefficient of ground friction is needed?

$$\mu = \frac{f}{N} = \frac{101 \text{ lb}}{181 \text{ lb}} = 0.558, \text{ a relatively low coefficient}$$

Given the mechanical advantages of the brake lever systems (section 11.14), a stopping force of 101 lb is not difficult to produce; if you brake too hard, you can easily flip the bike into a header.

In a ride downhill, the upward pointing force of friction is increased by the need to counter the component of the weight parallel to the hill. Consequently there is an increased danger of a header that develops when riding down a hill.

Why do rear brakes lose effectiveness during a hard stop?

During deceleration, there is shifting of weight from the back tire to the front tire. As the rear normal force is reduced, the frictional stopping force is also reduced. The rear brakes lose effectiveness during quicker stops. This reduced effectiveness is especially important on bikes equipped with only rear coaster brakes, common on many American cruiser bikes and children's bikes. The author has also observed poorly designed children's bikes, from well-known manufacturers, that have a single hand brake that operates on the rear wheel. This is the worst possible combination of a child's low hand strength braking with a wheel that loses contact with the ground during a hard stop.

Is there any way to safely demonstrate a 'header' without danger of going over the handlebars?

There is a very simple and safe way to demonstrate the physics of the header. With the bike at rest, stand with your feet on the ground and straddle the bike's top tube. Squeeze hard on the rear brake lever and 'lock' the rear wheel. Simultaneously push the bike forward by the handlebars; you will observe that, although the rear wheel is locked, the bike merely slides along the ground. The rear wheel does not induce a header since the braking force is lost as soon as the wheel lifts ever so slightly off the ground. If you repeat the experiment and lock the front tire, the bike will not slide. The bike acts as if the front wheel is 'glued' down. The rear tire rises up and the system rotates about the front tire's ground contact.

References

[1] Velox 1869 *Velocipedes, Bicycles, and Tricycles* (New York: Routledge) p 52

[2] Berto F 2009 *The Dancing Chain—the History and Development of the Derailleur Bicycle* 3rd edn (San Francisco: Van der Plas) pp 375–377

[3] Gunther C M *et al* 2008 Grip strength in healthy Caucasian adults: reference values *J. Hand. Surg. Am.* **33** 558–65

[4] Wilson D G 2004 *Bicycling Science* 3rd edn (Cambridge, MA: MIT Press) p 244

Chapter 12

Centripetal acceleration—turning and bicycle stability

Finally, in this last chapter, we comprehend how the simple and beautiful laws of physics harmonize and progenerate the magical charm of the bicycle's balance. The mystery of the bike is unraveled as an imperceptible interplay of motion and the forces of gravity and friction. The physical concepts work their enchantment unbeknownst to the rider.

What makes the bike turn? How do the internal forces of the rider on the bike create the necessary external force to change a bike's direction?

12.1 Review of Newton's laws—centripetal force and acceleration

Before examining the motion of an object executing circular motion, we pause to recall that Newton's First Law of Motion (chapter 5) tells us that an object at rest or traveling in a straight line will continue to do so unless acted upon by a *net external force*. From Newton's Second Law, we saw that the net force produces a change of velocity, an *acceleration*, in the direction of the force. Most of our acceleration examples were objects traveling in straight lines, e.g. a bike starting from rest, speeding up, or slowing down. These accelerations resulted in a change in the object's speed, the magnitude of the velocity vector.

It is important to keep in mind that the definition of a vector involves both the *magnitude* and the *direction* of the vector. An object traveling at a constant speed but undergoing a change in direction will be accelerating. If you are walking down the street and make a left turn at an intersection, you underwent an acceleration! The acceleration was associated with the change in your direction. The acceleration to the left requires a force to the left. Take a walk down the street and carefully observe your actions during this left turn. You push on the ground to the right and, from Newton's Third Law, your rightward push on the ground results in the ground

doi:10.1088/978-1-6817-4441-4ch12

pushing on you toward the left. This force from the ground is the net external force that accelerates your velocity with a change in the vector's direction. There does not have to be a change in the vector's magnitude; your speed can be constant during this entire walking and turning process.

As another example, envision riding your bike at a constant speed around a circular racetrack. The magnitude of your velocity does not change; however, the direction of your velocity constantly changes. Suppose you are going counter-clockwise around the track; at one point you are heading north; an instant later you are going northwest; eventually, your bearing is west. Round and round you go with a continuous change in direction. *The rider and bicycle are continuously accelerating without any gain or loss of speed. The acceleration is a consequence of the change in the velocity's direction rather than its magnitude.*

From the work of Galileo and Newton, we learned that accelerations are produced by *net external forces*. The net external force and the acceleration are both vectors and will be in the same direction. Gravity, ground contact forces, and air resistance are the usual external forces to the bike/rider system. Forces between the rider and the bike and various components of the bike such as pedals, handlebars, and seat are not external forces; rather, they are internal forces within the system. *Internal forces cannot accelerate* the motion of the system.

Another important aspect of motion is inertia. *Inertia* is the tendency of an object to travel at constant velocity—in a straight line at a constant speed. The *mass* of an object is a measure of its inertia. Newton's First Law of Motion states:

An object at rest will remain at rest and an object in motion in a straight line will remain in motion in a straight line unless acted upon by a net external force.

The implication of Newton's laws is that, when an object is traveling in a circle, it must be continuously acted upon by a net external force. If this net external force is removed, the object's inertia causes the body to travel in a straight line. Common examples are the circular motion of the Moon orbiting the Earth, a bucket of water swung in a circle, a rock whirled on a string, and *a cyclist making a turn*. In chapter 10, we used the situation of whirling a rock on a string as an example of circular motion (figure 10.4). As we think about this rock's velocity, we see it is constantly accelerating. It is not speeding up, being whirled faster and faster; rather, the rock is accelerating because of the velocity's change in direction. The net external force on the rock is the tension in the string; if the string breaks, the rock's inertia causes it to fly off in a straight line.

To obtain an expression for the acceleration, a bit of intuition is needed—imagine you are the one holding the end of the string. Instead of just whirling the rock at a constant speed as discussed above, you now whirl the rock faster and faster. You will notice that it is necessary to exert an increasingly larger force pulling inward on the rope. The acceleration and required force on the rock is proportional to the square of the rock's speed. You experience the same effect if you shorten the radius of the rock's orbit. The acceleration and force are inversely proportional to the orbit radius. The acceleration is written:

$$a = \frac{v^2}{r} \tag{12.1}$$

and the force is:

$$F_c = ma_c = m\frac{v^2}{r} \tag{12.2}$$

The acceleration and force *are both directed* toward your hand at the center of the circle! Traditionally, the adjective *centripetal* is used to describe the acceleration and the force. The word centripetal comes from a union of Latin words; it means 'to go to the center'. The *centripetal acceleration* is pointing *toward the center of the turn*—as the object travels in its circular path, it is accelerating toward the center. A net external force, appropriately named the *centripetal force*, produces this acceleration. A subscript c indicates centripetal quantities.

Consider the circular motion of an object as being comprised of two independent elements:

- *Inertia*—the tendency to travel in a straight line with constant velocity (Newton's First Law).
- *Acceleration*—the change in the object's velocity toward the center of the circle.

What is the source of these forces pulling a body toward the center? A few examples:

- In the case of the Moon orbiting the Earth, the centripetal force is the Earth's gravity; the Earth pulls on the Moon just the Earth pulls on you.
- In the case of a rock whirled on a string, the centripetal force is the tension in the string. If the string breaks, the rock flies off in a straight line tangent to the circle.
- In the case of a bicycle making a turn on a level surface, the centripetal force is the force of friction from the ground.
- In the case of a bicycle making a turn on a banked road, the centripetal force is a component of the road's normal force.

The concept of circular motion is the most difficult to grasp in all of basic physical science. For the student and instructor, a difficult topic has been made even more challenging by the use of the terms *centrifugal acceleration* and *centrifugal force*. Although centrifugal concepts have valid physical meaning in certain studies of motion, they are always misused in elementary explanations. Unfortunately, generations of teachers and their students have been confused by the term. It is commonly stated that objects, moving in a circular path, will fly off in a straight line due to the centrifugal force. We hear that, during a washer's spin cycle, the water is forced out of the wet clothes by centrifugal force. Or, the passenger in the turning car was pushed into the door by the centrifugal force. Even worse, we find scientific instruments called centrifuges! Let us put this concept to rest—at least in this book. There is no force causing the water in wet clothes to fly outward. There is no force causing the passenger to be pushed into the door of the turning car. Centrifuges do not exert centrifugal forces.

Keep it simple, in basic physical explanations: *there are no centrifugal forces!!!*

Why then does the water get spun out of the clothes; what causes the passenger to be pressed into the door? It is inertia, the tendency for a body to travel in a straight line.

Objects, traveling along a curved path, are accelerating due to the change in their velocity vector's direction. This acceleration requires a force; should the force be insufficient (maybe you let go of the rope or hit a patch of ice when turning on a bike), the body moves in a straight line due to inertia.

12.2 Making a turn

What is the source of the centripetal force that causes a wheeled vehicle to turn?

For a bicycle or car on a level (not banked) surface, the centripetal force is friction between the tires and the road. Since the centripetal force must point toward the center of the turn, the force is horizontal.

It is best to explore the role of friction with a simple example—a tricycle ridden by a young child on level ground; we use a tricycle to avoid issues of balancing. Although precocious, our youngster has not yet studied physics in preschool; she simply pedals merrily along. The diagrams in figure 12.1 show the horizontal ground forces on the wheels. Initially, the child is traveling straight (figure 12.1A).

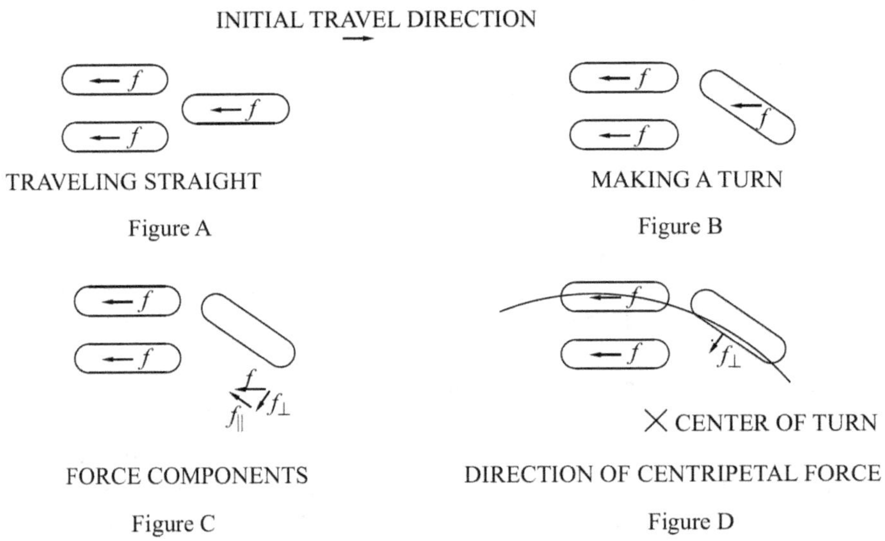

Figure 12.1. (A,B,C,D) Tricycle wheel.

The horizontal ground forces on the rolling tricycle are simple—rolling resistance and static friction with the ground. With a rigid wheel and hard ground, rolling resistance is nil. The force of static friction serves to keep the wheels spinning (chapter 5) but causes no loss of speed since there is no work involved. Even a two-year-old knows the actions required to steer—she turns the handlebar and redirects the front tire (figure 12.1B). The youngster's force on the handlebar grips is internal within the tricycle/rider system.

The road forces on the two rear wheels are unchanged and can be ignored; notice, the front wheel is no longer aligned with the direction of travel. Since the road force is a frictional force, it will oppose the motion. The force's direction is opposite to the velocity vector. With the force of friction at an angle to the front wheel, the force can be resolved into components parallel and perpendicular to the wheel (figure 12.1C). The component parallel to the wheel $f_{\parallel}$ behaves similarly to the rear wheel forces. The component perpendicular to the wheel $f_{\perp}$ is the force responsible for changing the direction of the tricycle. This component of static friction, between the wheel and the ground, is the centripetal force that produces the centripetal acceleration toward the center of the curvature (figure 12.1D).

The events, described above for a kiddie on a tricycle, apply to many other wheeled vehicles such as bicycles and autos. Let us look into the parameters for the turn, using the expression for the centripetal force:

$$F_c = m\frac{v^2}{r}$$

And the force of friction:

$$f = \mu N = \mu mg \tag{12.3}$$

As the centripetal force:

$$\mu mg = m\frac{v^2}{r}$$

The mass cancels out and we solve for the speed:

$$v = \sqrt{\mu gr} \tag{12.4}$$

How fast can she make a turn of a tight radius of 25 ft?

- with a high coefficient of friction between the tires and the road, e.g. $\mu = 0.9$

$$v = \sqrt{\mu gr} = \sqrt{0.9 \times 32.2 \text{ ft/s} \times 25 \text{ ft}} = 26.9 \text{ ft/s} = 18.3 \text{ mph}$$

- on a patch of ice with low friction $\mu = 0.1$

$$v = \sqrt{\mu gr} = \sqrt{0.1 \times 32.2 \text{ ft/s} \times 25 \text{ ft}} = 8.97 \text{ ft/s} = 6.1 \text{ mph}$$

The ability to turn is independent of the body's mass. Of course, in a situation of zero friction, there is no centripetal force and the body continues in a straight line path.

12.3 Banked surface

Why is it better to have a 'banked' road when making a turn?

A banked roadway helps provide the centripetal force needed to make a turn. Consider figure 12.2A, which illustrates the rear view of a rider making a right hand turn on a banked surface. Suppose that the surface makes an angle θ with the horizontal. Notice, the road is banked toward the inside of the turn; the upward push from the road is angled toward the center of turn.

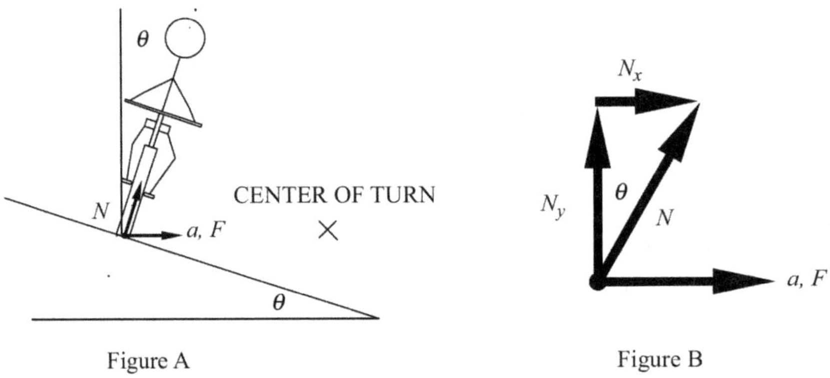

Figure 12.2. (A,B) Banking of road.

The normal force of the road is not vertical and, therefore, has a component pointing toward the center of the turn N_x; this *horizontal component from the ground functions as a centripetal force*. In most situations, this force acts in concert with friction but, if the banking angle is sufficiently large, the role of friction is minimized.

12.4 Equilibrium and stability

Throughout this book, we looked at many situations in which the net external force is zero (chapters 5 and 6). When the external forces sum to zero, the object has no linear acceleration; it is described as being in *translational equilibrium*. We have also explored situations in which the net external torque is zero (chapter 10); the object has no rotational acceleration; it is said to be in *rotational equilibrium*.

Objects in a condition of translational and rotational equilibrium are referred to as *stable*; stable objects are *balanced*, i.e. they will not fall over. However, as we will see shortly, it is possible for an object to possess neither translational nor rotational equilibrium but yet be 'stable' and not fall to the ground. The next few sections explore common conditions of stability in both equilibrium and non-equilibrium situations.

12.5 Equilibrium and stability with multiple points of support

Consider a body supported at multiple points, for example a table.

The force of gravity on the table is considered acting at a single point—the center of mass or *center of gravity*. If the table sits on a level floor, the downward force of

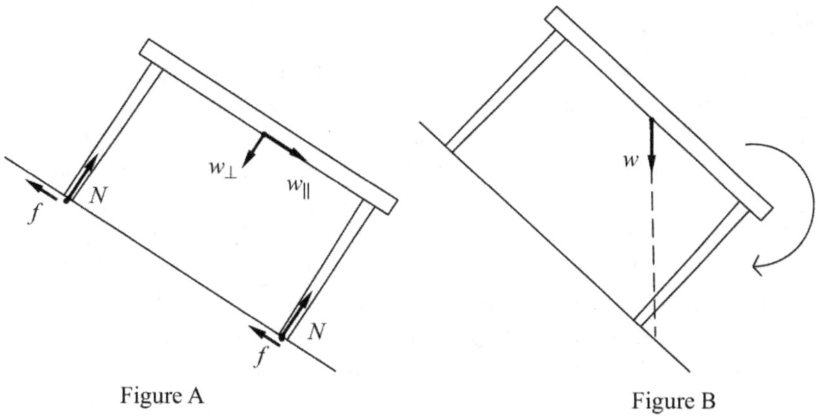

Figure 12.3. (A,B) Stability of table.

weight and upward normal forces from the floor sum to zero and there is no linear acceleration; also, there are no unbalanced torques and, hence, no rotational acceleration. The table is in equilibrium.

What happens if the table is sitting on a sloped surface (figure 12.3A)? Initially, consider a small slope of 10 °. The force of gravity now has a component parallel $w_{||}$ to the slope (section 6.4) but, if there is sufficient friction, the table remains in equilibrium and does not slide down the hill. The table is in translational equilibrium since the forces sum to zero. If the hill is steep enough, or if the coefficient of friction is too small, the upward force of friction might not match the downward component of gravity and the table accelerates down the hill. The table is not in translational equilibrium since the forces do not sum to zero.

Concern for an object perched on a steep slope does not end with the possibility of the object sliding down the hill. Another, and perhaps more disastrous, result, is having the table topple over. Even when the friction is high, the table can tip; it can be rotationally *unstable*. Figure 12.3B shows the weight and its line of action on a steep slope. There is an important change in the line of action of the gravitational force—it is now to the right of the lower legs; the weight produces a clockwise torque and the table topples. A stable table requires that the weight's line of action falls between the legs' contact with the floor. A tall narrow table with closely spaced legs will topple more readily than a low wide table.

We have an easy and intuitive method of determining an object's rotational stability: does a vertical line drawn from the object's center of gravity pass between the points of support? If so, the object is rotationally stable. Most tables are very stable—what of other objects? A human standing on two feet will be stable. Stand on one foot—not so stable. A body with a single point of support located below the center of gravity will be in delicate balance. In the next section, we examine the balance of objects with a single point of support.

12.6 Stability of runners

Before moving to a discussion of the bicycle's stability, we offer for consideration another joyous sight—a little kid, maybe two years old, running round and round in tight circles. In addition to being fun to watch, the child demonstrates an example of a body stable and in balance under conditions of translational and rotational non-equilibrium. Observe carefully as the youngster runs at top speed in a tight (maybe 10 ft) circle. You will see that the child is leaning toward the center of the circle (figure 12.4A). No doubt, toddlers are smart and, while they have not yet studied physics, they intuitively understand the basic principles. Running at a constant speed in a circle involves centripetal forces and centripetal accelerations. To balance in this condition of translational non-equilibrium, her body must also be in rotational non-equilibrium—*she must lean in the direction of the acceleration*—toward the center of the circle!

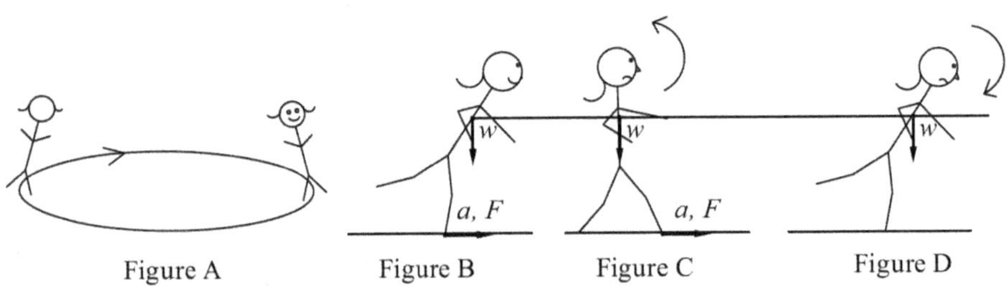

Figure 12.4. (A,B,C,D) Lean of kid and sprinter.

This simple, easy to observe, example is a perfect model for the marvel of balancing on a bicycle. We can learn a lot from these little kids!

12.7 Stability of sprinter

Use your legs, take the start, run away.

Merchant of Venice,ii, Shakespeare

Perhaps our little two-year-old bundle of energy has progressed a decade or two later to a medal winning sprinter. Sprinters exhibit a phenomenon that is very easy to observe. When a sprinter accelerates from the start block, her body must lean forward as shown in figure 12.4B. Her center of gravity is forward of the leading foot. Since she is accelerating, her body is *not in translational equilibrium* and since her center of gravity is not above her single point of support, she is *not in rotational equilibrium*!

In spite of the lack of either translational equilibrium or rotational equilibrium, the sprinter is very stable; she does not fall to the ground. To gain an intuitive understanding of this situation, consider what would occur if the sprinter did not lean. Suppose she stands straight up and begins to run (figure 12.4C). As her feet accelerate to the right, inertia causes her center of mass to be left behind and she falls backwards (counterclockwise).

What if she did the opposite—leaned forward and did not accelerate (figure 12.4D)? Her body would rotate clockwise and she would topple over. In the normal technique for sprinting, she leans forward at the same time her feet accelerate. Her stability is a consequence of the linear acceleration of her feet producing a counter-clockwise rotation, and an angular acceleration of her overall body creating a clockwise rotation. Intuitively, we can consider the two rotations as canceling.

The need to lean ceases when the runner reaches top speed and continues at a constant velocity.

12.8 Equilibrium and stability with single point of support

Is it possible to support an object with a single support point?

It is easy to support and balance an object when the single point of support is above the center of gravity and you do not have to be a track star; all you need is a broom or umbrella. Figure 12.5A illustrates a broom held with the heavy section (center of mass and gravity) in the lower position.

When the broom is held as in figure 12.5A, it is in both translational and rotational equilibrium. If the broom is subjected to sideways displacement, it returns to the stable position. 'Stable equilibrium' describes a body supported from above its center of mass.

What happens when the broom is held upside down as shown in the next figure (figure 12.5B)? To avoid accidently applying a torque with the hand, we must keep the palm open and not grip the handle. The broom in figure 12.5B is precarious. Objects with the center of gravity above the point of support are in a condition of 'unstable equilibrium'. Even when balanced on a rigid support such as the floor, the object is easily toppled by small disturbances (breezes, vibrations) that slightly displace the center of gravity to the side of the support point. The result is that the force of gravity creates a torque about the support point and the broom's angular acceleration rotates it into a fall (figure 12.5C).

Is it not simple to balance a broom in the upside-down position? Yes—with a bit of practice. As long as the hand is free to move, we can keep the broom from falling

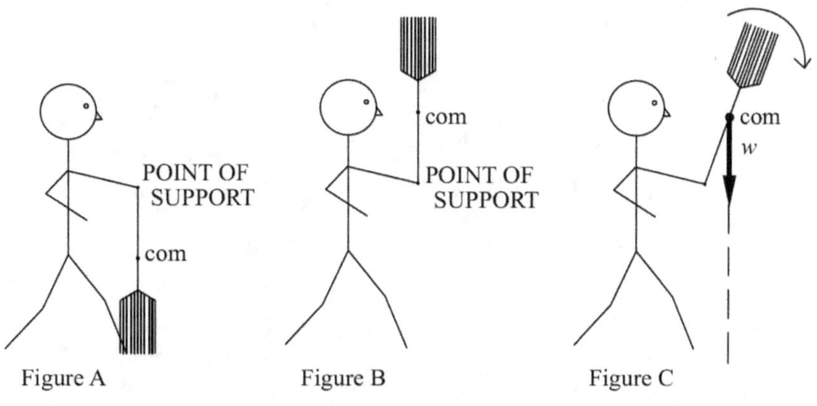

Figure 12.5. (A, B, C) Stability of brooms.

by continuously repositioning the hand left and right, to and fro under the broom's center of gravity. Stamina is the primary limitation. However, the broom will be *balanced* over the point of support for only a small percentage of the time; most of the time the broom will be 'falling' one way or the other. If the hand is restricted in its movement, the broom falls.

12.9 Stability of broom when not in equilibrium

The most common way to keep the broom from falling is to constantly shift the hand—as discussed in the last paragraph. Nevertheless, there *is another way to 'balance' the broom*—especially for the swift footed. It is possible to travel with the broom in a stable but non-equilibrium position by imitating the sprinter. If the broom leans past its point of support and accelerates in the direction of the lean, the broom will not fall (figure 12.6A).

In order to maintain this maneuver, you need lots of room and swift legs. Actually, it is not your speed that matters; rather, it is your ability to maintain the acceleration. The forces on a leaning broom are shown in figure 12.6B:

N—the upward force exerted by hand.

f—the horizontal force of friction between hand and bottom of broom.

w—weight of broom, shown at its center of mass (COM).

We analyze the motion of the broom as independent, separate actions: translational motion across the room and rotational motion about the broom's center of mass.

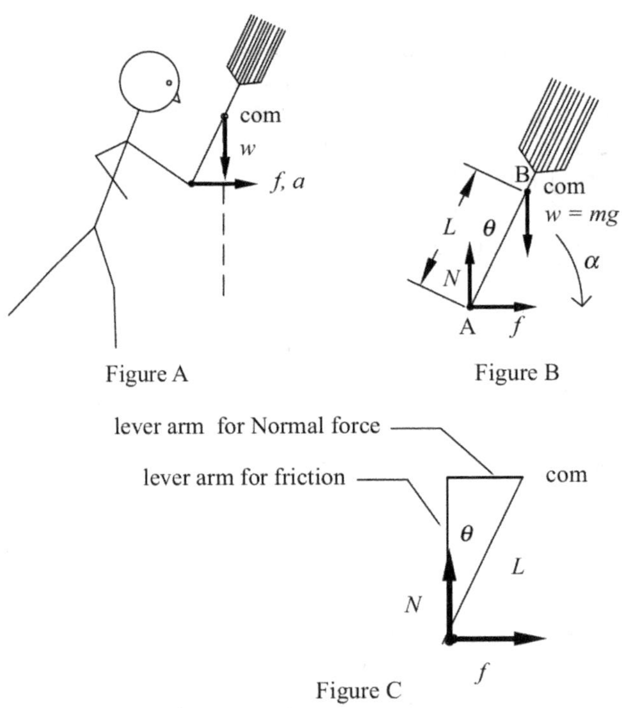

Figure 12.6. (A,B,C) Accelerating brooms.

The horizontal translation of the center of mass is determined by summation of the forces, a straightforward application of Newton's Second Law of Motion. The only horizontal force is the frictional force exerted by the open hand:

$$f = ma \tag{12.5}$$

If the broom is moved at constant height, there is no vertical acceleration and the vertical forces sum to zero. The upward force from the hand N, equals the downward pull of gravity w:

$$N = w$$

In section 6.2 (equation (6.2)), we developed the expression:

$$w = mg$$

or

$$N = mg \tag{12.6}$$

The broom's rotation is determined by the net torque, calculated about the center of mass (point B). L is the distance along the broom from the base (point A) to the COM at point B, *the angle of the broom's lean is* θ.

Force N with a lever arm of $L \sin\theta$ produces a clockwise torque about the COM; force f with its lever arm of $L \cos\theta$ produces a counterclockwise torque. If the broom is not 'falling over', its rotational acceleration α is zero and the torques must sum to zero:

$$\begin{aligned} \tau_{\text{cw}} &= \tau_{\text{ccw}} \\ NL \sin\theta &= fL \cos\theta \end{aligned} \tag{12.7}$$

or after eliminating L: $N \sin\theta = f \cos\theta$

$$\begin{aligned} \frac{\sin\theta}{\cos\theta} &= \tan\theta = \frac{f}{N} = \frac{ma}{mg} = \frac{a}{g} \\ \tan\theta &= \frac{a}{g} \end{aligned} \tag{12.8}$$

We have an important relationship between the acceleration of the broom and the angle at which it must lean. For example, if you wish to accelerate this broom at 10 ft/s^2, the broom must be leaning at an angle of:

$$\tan\theta = \frac{a}{g} = \frac{10 \text{ ft/s}^2}{32.2 \text{ ft/s}^2} = 0.311$$

looking up the angle in appendix C or using a calculator: $\theta = \tan^{-1} 0.311 = 17.3°$.

While a broom leaning at 17.3° is not a problem, the acceleration of 10 ft/s^2 might offer a challenge. After just three seconds, your speed will be 30 ft/s, near 20 mph!

Thinking back to the sprinter—the same principles and results apply. Given a particular acceleration out of the starting block, equation (12.8) can be used to obtain her required lean angle. The physics and mathematical analysis of the broom's lean and acceleration are identical to the sprinter. We can intuitively view the stability of these as a combination of a linear acceleration at the bottom and an acceleration of the COM caused by the rotation of the boom.

Note: we have described the only way to move a broom in a continuous, steady motion across the room. You cannot move the broom at a constant velocity—only under conditions of constant acceleration. It might be possible to move in a semi-steady fashion through a series of quick jerks back and forth, i.e. we make it 'fall' one way and then quickly move the bottom of the broom ahead and make it fall backward. By continuously repeating these actions, the broom could translate across the room. The motion, however, is not a steady forward movement.

In considering the motion of the sprinter and the broom, there are three important concepts that have direct application to a turning bicycle. They are:

- The broom's COM must be leaning past the point of support. The broom is not in rotational equilibrium.
- The base of the support must be continuously accelerated. The broom is not in translational equilibrium. If the acceleration ceases, the broom falls over!
- A force of friction provides the accelerating force.

There is another aspect to the broom's motion that is common to that of the bicycle; it is a *countermove*. Before beginning the acceleration of the broom to the right, you must make a quick move to the left. Inertia, the tendency of the broom's mass to stay at rest, causes the COM to remain in the original location and the broomstick is now leaning toward the right. What transpires if you do not make the quick countermove to the left? Suppose you begin with acceleration of the broom's bottom to the right. Try it! The broom will be leaning left and you had better be prepared to accelerate to the left.

12.10 Stability of bicycle when not in equilibrium

What makes the bike balance?

The actions of riding and balancing a bicycle are very similar to the balancing of runners and brooms. In the previous few sections, we discussed the stability of a sprinter, a broom, and an enthusiastic kid. All of these examples were in neither translational equilibrium nor rotational equilibrium. Yet, despite this lack of equilibrium, the sprinter, the broom, and the kid are very stable; they do not fall down! They are balanced in spite of being in states of non-equilibrium. *Their stability is a consequence of the fact that they simultaneously lack both translational and rotational equilibrium.* These considerations are very important in understanding the stability—'balancing'—on a bicycle.

Why must we lean while making a turn on a bicycle?

The difference between the bicycle and circling toddler (figure 12.4A) compared to the sprinter and broom is that the sprinter and broom are undergoing linear accelerations, whereas the turning bike and happy kid are undergoing centripetal accelerations toward the center of the turn. Just as the sprinter and broom must lean in the direction of the acceleration, so must the bicycle and kid lean in the direction of the acceleration, toward the center of the circle. The bike and the kid running in circles are most similar in that they lean as they undergo a centripetal acceleration toward the center of the circle!

Once again, we see a connection between bikes and kids—it is baked into physics!!!!

The view from behind our rider, as he is making a right hand turn, illustrates the lean toward the center of the turn (figure 12.7A). The reasoning, developed for the broom in is perfectly applicable to the bicycle system. (Compare figures 12.6B and 12.7B and substitute the head of the rider for the broom bristles).

Section 12.9 gave an expression for the accelerated broom's angle of lean. The rider/bike geometry is shown in figure 12.7B. The system center of mass is somewhere in the middle of the rider's body. There is no vertical acceleration of the system; the COM stays at the same height. If the road is not banked, all of the centripetal force is due to friction. The analysis is exactly like that of the leaning broom; the result is the same condition on the angle of the lean:

$$\tan\theta = \frac{a}{g} \tag{12.9}$$

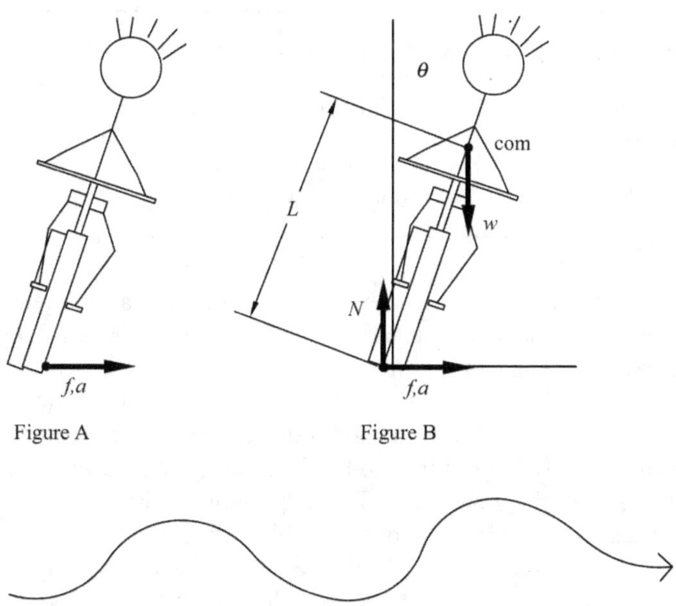

Figure 12.7. (A,B,C) Geometry of lean.

The important difference between the broom and the bicycle is that the broom must undergo a *linear acceleration—across the room*; the bicycle's acceleration is a *centripetal acceleration into the center of the turn*! From section 12.1, the expression for centripetal acceleration is:

$$a = \frac{v^2}{r} \tag{12.10}$$

Combining the above two equations:

$$\tan\theta = \frac{v^2}{gr} \tag{12.11}$$

This expression gives the required lean angle for a given speed and radius of turn. Notice, the lean angle does not involve the system's mass; it is a good thing that a heavily loaded bike will have the same lean angle as one lightly loaded. Otherwise, the adage 'once you learn how to ride a bike, you never forget' might not be true as our weight increases with age. Perhaps, even more important is that the lean angle of a very heavy bike system—such as a tandem—is identical to that of a solo machine.

How is it possible for the rider to know the exact angle at which to lean the bike?

Any errors in the rider's angle of lean θ are easily compensated by steering into a turn of slightly different radius; she adjusts the radius of the turn r such that the above condition (equation (12.11)) is met.

Thus, we have the secret to balancing a bike.

> The two-wheeler is always in a state of unstable equilibrium with its points of support—the tires' contact with the ground—below the center of mass. Once the rider lifts her feet off the ground and begins to travel, the machine leans to one side. In order not to fall over, the bike must be steered in the direction of the lean and, with the bike now executing a turn, has a centripetal acceleration. The combination of the lean and centripetal acceleration results in a condition of balance.

What occurs next is mostly up to the rider; she might change the combination of turn radius, speed, and lean angle. Alternatively, she could shift her weight; or, perhaps, wind causes a lean to the opposite side. She now makes a subtle adjustment in steering direction and executes a turn in the new direction of the lean. While our cyclist may be trying to ride in a straight line, careful observation will reveal the actions of leaning one way, steering into the lean, and then leaning the opposite way and now steering into the new lean. The experienced rider repeats these actions continuously and subconsciously. What might be considered a ride in a straight line is actually a series of turns of very large radii; figure 12.7C is a top view of a rider, trying to travel in a straight line but, nevertheless, tracing out a serpentine path.

Life is like riding a bicycle. To keep your balance you must keep moving.
Dr Albert Einstein

For the individual leaning to ride, the challenge is realizing that the bike must be moving for it to balance; it is not possible to execute circular motion without the bike being in motion. The learner must also develop a knack for turning into the direction of the falling bike's lean. In a later section, we will see that bicycles possess a self-stability owing to the front fork design; a leaning bicycle will automatically steer into a turn. The neophyte must learn not to fight the bike's self-steering action.

If you can force your heart and nerve and sinew
To serve your turn long after they are gone,
And so hold on when there is nothing in you
Except the Will which says to them: "Hold on!"

If by Rudyard Kipling

A few specific calculations, using equation (12.11), illustrate the balance condition. Suppose we have a rider/bike system of 181 lb (5.62 slugs) making a turn at 20 mph (29.3 ft/s) in a circle of radius 100 feet.

First, solve for the angle of lean:

$$\tan\theta = \frac{v^2}{gr} = \frac{(29.3 \text{ ft/s})^2}{32.2 \text{ ft/s}^2 \times 100 \text{ ft}} = 0.267,\ \theta = \tan^{-1} 0.267 = 14.9°$$

Solving for the magnitude of the acceleration:

$$a = \frac{v^2}{r} = \frac{(29.3 \text{ ft/s})^2}{100 \text{ ft}} = 8.59 \text{ ft/s}^2 = 0.267 \text{ ‘g’}$$

We can also get the necessary force of friction:

$$f = ma = 5.62 \text{ slugs} \times 8.59 \text{ ft/s}^2 = 48.3 \text{ lbs}$$

And the required minimum coefficient of static friction:

$$\mu_{\text{min}} = \frac{f}{N} = \frac{48.2 \text{ lb}}{181 \text{ lb}} = 0.267$$

The good news is that this coefficient is well below the coefficient for rubber-road contact of 0.9 (table 5.1).

The results of these calculations and other combinations of speed and turning radius are shown in table 12.1.

The bottom two rows in table 12.1 involve extreme leans. Can this be done? We are not sure of 40 mph turns by your average Sunday afternoon cyclist, but bicycle and motorcycle racers routinely perform high-speed turns. The balancing and turning of motorcycles are subject to the same laws of physics as bicycles. Perhaps, the reader has observed a little trick used by high-speed racers—they extend their knee toward the inside of the turn. The leans are so extreme for the motorcyclists that they wear kneepads. In our calculations on the system center of

Table 12.1. Parameters when making a turn.

v	r (ft)	a (ft/s^2)		θ	f(lb)	μ_{min}
20 mph (29.3 ft/s)	100	8.59	0.267 'g'	14.9°	48.3	0.267
20 mph (29.3 ft/s)	50	17.2	0.533 'g'	28.1°	96.7	0.533
40 mph (58.6 ft/s)	100	34.3	1.07 'g'	46.9°	192.8	1.07
40 mph (58.6 ft/s)	50	68.7	2.13 'g'	64.8°	386.2	2.13

mass, we normally envision a (head to toe) symmetrical bike and rider as seen from behind; by extending the knee toward the inside of the turn, the rider is able to move the center of mass off the line of symmetry. This maneuver results in the machine maintaining a more vertical orientation providing road contact closer to the tire's center.

The other issue with the 40 mph speeds is that the friction coefficients are high; a banked road is necessary.

Why do riders fall over if they make a turn that is too fast and sharp for ground conditions?

Look carefully at the forces on tire bottom (figure 12.7B); the force of friction pointing to the right is keeping the bottom of the wheel from kicking out to the left. This force, producing a counterclockwise torque about the COM, opposes the clockwise torque produced by the normal force. If, in the course of making a turn, you hit an area of low friction (a wet section of road or loose gravel), the force of friction is insufficient and the bike rotates clockwise slamming the rider sideways onto the ground. The author speaks from experience on this experiment.

12.11 Self stability of a bicycle

What is meant when people say a bicycle possesses 'self stability'?

We have seen that a turning bicycle must lean; conversely, a leaning bike must turn. Since a bike/rider system has a center of mass perched far above the road support, the machine readily leans to one side or the other. Much of the skill in learning to ride is to accept the lean and steer into a turn. A lean to the opposite side requires a steer into the new direction. A cyclist can ride for hours subtly executing the proper combinations of leans and turns. The skills become intuitive and the actions are subconscious. To get a feeling for these delicate and continuous actions, try to ride along a painted road line. No matter how hard you strive, the bike wanders from one side of the line to the other.

In addition to intuitive skills and a deep-rooted aversion to falls, the rider is facilitated by important features in the machine's design. Aspects in the fork and front wheel design afford the bicycle a significant property of *self-stability*. The self-stability allows even a riderless bicycle to travel and balance in an astonishing manner. The bicycle's stability has been the subject of much speculation, scientific research, and analysis. It is also common to find erroneous explanations of the stability. Serious efforts have developed models and theories that quantify and allow for predictive evaluations of the machine's stability. The interested reader may wish to explore the scientific literature to gain an appreciation for the complexity of the concepts, and the level of mathematical sophistication that is brought to a rigorous analysis [1]. Since the basic approach of this book is to understand the physical behavior of the bicycle with minimum use of mathematics, we offer simplified, intuitive, explanations of the bicycle's self-stability. Most bicycle models are based on frame, fork, and wheel configurations that originated in the nineteenth century. These early designs, likely evolved from empirical methods of trial and error, persist in modern machines.

You will never plow a field by turning it over in your mind.

Irish Proverb

In order to grasp many aspects of the physical world, it is best to use the *experimental method*—your own personal powers of observation—look, measure, think, look again; do your own experiments. In the next few paragraphs, we discuss simple experiments that illustrate important aspects of the bike's stability.

Experiment #1: instability caused by a locked fork

Any good science experiment needs a control—an experiment in which the event of interest does not occur. Is there a way to eliminate or reduce the bike's stability? Yes—it is very easy. If the front fork is prevented from freely turning, such as by removing the upper and lower headtube bearings and tightening the locking nuts, the bicycle with a locked front wheel will lean and fall as soon as it begins moving. Another method for locking the fork is to brace the handlebar/fork/wheel system in a forward position and drill a hole straight through the headtube and fork. Lock the fork by inserting the drill bit through the holes; removal of the bit unlocks the steering mechanism.

A bike with a locked front fork will quickly fall—with or without the rider! If you try to ride this bicycle—*be forewarned: you will not be able to 'balance' the machine regardless of how fast you pedal and spin the wheels*! The bicycle will fall over in a few feet. Wear a helmet; keep the saddle low, and do not clip into pedals!

The stability of a bike is inherently dependent on the free rotation of the front fork.

Experiment #2: stability of a riderless bicycle

The reader is encouraged to demonstrate the self-balancing phenomena of a riderless bike for herself. All it takes is a bike with a straight frame, good air in the tires, and a front fork that is free to turn. Unless someone catches the bicycle, it will eventually crash. Best to use an inexpensive bike; most ten-dollar flea market finds will work fine. Also, since a riderless bicycle is capable of traveling far (miles?), be certain that the bicycle does not encounter objects.

The ideal location for experimenting with a riderless bike is an open area with a gentle downward slope. Run alongside the bicycle and give a moderate push on the saddle; do not jerk or twist the handlebars. The slope will ensure that the machine maintains its forward speed. As the bike travels away, it begins falling to one side. Observation is the key to a good experiment; notice, as the bicycle leans to one side, the front wheel steers into the lean; the bicycle now moves along an arc and then quickly rights itself back up. Next, the machine then leans in the opposite direction and the sequence is repeated. The process can occur many times. It is quite marvelous to see the riderless bike travel long distances—the primary limitation being when the two-wheeler loses speed and cannot right itself from a lean.

These experiments show that the bike with a front fork, free to rotate, has remarkable self-stability; it is able to 'balance' itself and travel long distances with no rider control. It is worthwhile repeating the riderless bike experiment with the front fork locked—the bike falls immediately.

To understand the bicycle's stability, we need to look close at several key aspects of the front fork design.

12.11.1 Weight distribution of the handlebar– fork– front wheel

Before examining the bicycle, you can get a sense of the gravitational effect on the front fork by holding a heavy book (a twenty pound physics book is ideal) in your hand with the arm straight at an angle of about 45 ° to the vertical (figure 12.8A). Think of your fore and upper arms as the axis of rotation. A slight twist of the arm results in the weight being off to the side and gravity pulling the arm into further rotation. Now, instead of rotating your arm, lean your body to one side. Once again, the weight of the book causes the arm to rotate further. This example is a nice model for the behavior of the front fork design. The handlebar–fork–front wheel apparatus is designed with a center of mass forward of the fork's axis of rotation as it passes through the headtube. Inspection of the stem shows that the handlebar is forward of the axis and the wheel is also held forward of the axis by means of a sweeping curve at the bottom (figure 12.8B). Usually mountain bikes do not have a curved fork; rather, they employ a bend in the fork as it exits the headtube.

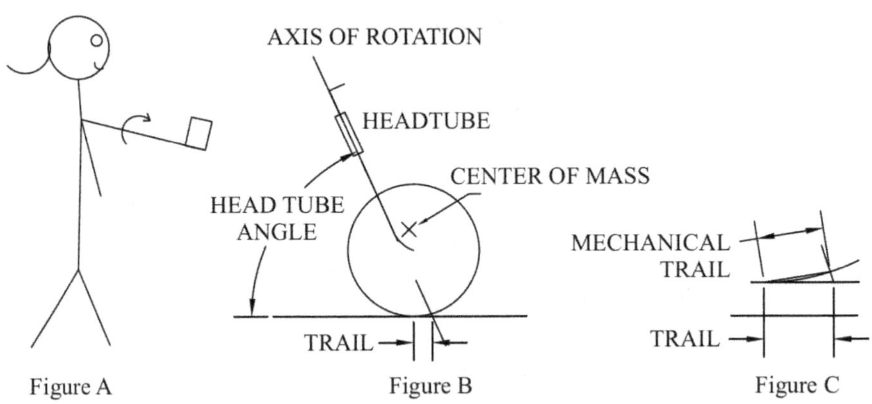

Figure 12.8. (A,B,C) Fork detail.

If you stand next to a bike, hands off the handlebar, and lean it to one side, the front wheel automatically turns into the direction of the lean. This rotation of the front wheel is not caused by any force from the ground; we can eliminate ground forces and still observe this effect. Get the front wheel off the ground; use a repair stand or even suspend the bike in the air with a couple of ropes attached to the top tube. When a bike leans, the front wheel automatically turns due to the gravitational pull on the fork's off-axis center of mass.

It is simple to observe the opposite effect when the bike is back on the ground. Just stand next to an upright bike and turn the handlebar; the bike leans in the direction of the turn.

Due to the interaction of the bicycle's components—a leaning of the bike causes the front wheel to turn. If the bike is moving, the leaning bike now travels along a circular arc. Conversely, if the rider deliberately turns the handlebars, the bike automatically leans in the direction of the turn. The turning/leaning bicycle meets the condition of stability/balance described in section 12.10.

12.11.2 Trail of the front wheel contact with the ground

The bicycle fork is angled forward; the angle of the headtube (typically about 72°) determines the angle of the fork. A straight line through the headtube is called the *steering axis*. The fork rotates about this steering axis line (figure 12.8B). The headtube angle is a specification given in a manufacturer's catalog. Note: to add confusion to a complicated topic, the headtube angle is also called the 'rake angle' or 'rake' of the fork.

Another experiment: sit on the bike's seat, lean forward, and look carefully from above the headtube toward the ground. You will see that the line of the steering axis projects to a point on the ground that is forward of where the tire makes ground contact. The front tire touches the ground a few inches behind the steering axis intersection with the ground (figure 12.8B). It helps to have an assistant mark these points on the ground. *The front wheel 'trails' the steering axis.* The *trail* of the bicycle is the distance between the projected point from the steering axis and the front tire contact. Careful inspection of figure 12.8B shows that the trail is a consequence of the slanting headtube/fork arrangement. Who first conceived such a sloped arrangement and why is not known; it may have been introduced as a scheme to position the handlebar closer to a seated rider. Whatever the purpose for its introduction, the sloped headtube/fork is the reason for the magical self-stability in the bicycle and contributes greatly to the ease with which the machine can be balanced and steered. Typically, the trail is a few inches; the trail of the bicycle is another design parameter given in a manufacturer's catalog.

Now, for an additional small complication; although the trail is specified, it is not the horizontal trail that matters. Rather, it is a component of the trail perpendicular to the steering axis, the so-called *mechanical trail*, that is the significant parameter (figure 12.8C). Since the headtube angles are large, the mathematical difference between the trail and its mechanical trail counterpart is only about 5%.

While the trail of the front wheel is a very important factor in the self-stability of the bicycle, one can sometimes have too much of a good thing. That is, too much trail will make the bicycle difficult to steer. The curved sweep at the bottom of the fork pushes the wheel forward and, therefore, reduces the amount of trail. In mountain bikes with suspension forks, the trail is reduced by a bend in the fork just below the headtube. Another experiment worth trying is to rotate the handlebars 180° (you might have to remove cables). With the sweep pointing backwards, the trail is greatly increased. The machine will be very stiff to steer.

How do the design of the bicycle and the laws of physics bring about this magical stability?

> The point at which the tire contacts the ground is significant; this is the location where ground friction acts and serves as the centripetal force on the machine. The trail of the front fork causes this centripetal force to act at a point behind the steering axis.

While still straddling the resting bike, lean it to the left and allow the handlebars to rotate toward the left. Since the bike is not moving, there will be no centripetal force; your helper can play the role of the centripetal force. Ask him to push with his finger on the right bottom side of the front tire (he would be pushing into the paper in figure 12.8B). Your helper's finger is acting as an artificial centripetal force. *The force of the push causes the front tire to become straight*; the push creates a torque about the bicycle's steering axis. The force's lever arm is the mechanical trail.

> In a turning bike, the centripetal force is applied by the ground friction to the bottom of the front wheel. The trail results in the centripetal force acting behind the steering axis and this force generates a torque that tries to straighten the front wheel. If the bicycle is turning toward the right, the centripetal force is also directed toward the right and, as it pushes on the bottom of the wheel, it tends to make the handlebars rotate counter-clockwise as observed by the rider. In order to maintain the right hand turn, the rider needs to apply a steady clockwise torque to the handlebars.

A turning wheel, mounted in such a manner that it tries to straighten out, is said to be castered. An automobile driver is keenly aware of this effect. If, during a turn, the hands are loosened from the steering wheel—the steering wheel returns to the straight ahead position. Thus, to maintain the turn, it is necessary for the driver to keep a steady torque on the steering wheel. Upon exiting the turn, the driver loosens her grip and allows the car's tires to straighten out. Supposedly, shopping carts employ castered wheels to ensure straight-line steering.

If you are ever lucky enough to find a shopping cart that actually steers well, check out the caster effect on the wheels.

There are several benefits from the caster effect.

One desirable feature of castering is that it encourages the vehicle, such as a bike or car, to travel in a straight line. When wind or road bumps push the vehicle to the side, the machine tends to straighten out and continue in a direct line. The role of the operator is to not resist the tendency of the machine to continue on its straight path. A second benefit of castering occurs when the operator desires to make a turn; castering offers a small resistance to the force of turning. This negative feedback requires that the operator maintain steering torque. The opposite, a positive feedback steering mechanism, will sharpen or 'take over' a turn; it would then be necessary to restrain the steering wheel or handlebar to avoid dangerous oversteering.

12.12 Summation of bicycle stability

Although the various factors, discussed above, act simultaneously in small, almost imperceptible increments to give the bicycle its stability, it is worthwhile to think of these actions in a step-by-step manner and get a feeling for the interplay of the factors. Consider, once again, the riderless bike traveling in a straight line. The rotations of the front fork are as viewed by an imaginary rider as she looks down from above the handlebars.

Step 1. the *bike begins to lean* to one side (right).

Step 2. the *front wheel rotates* (clockwise) into the lean. This action is the result of the front fork assembly having its center of mass in front of the steering axis; the gravitational pull on the unsupported mass causes the front wheel to turn in the direction of the lean.

Step 3. with the front wheel rotated, the *bike executes a turning arc to the right.* The radius of turn depends on the front wheel's rotation; with the riderless bike, the radius is usually large.

Step 4. the bike, leaning and moving along the circular path, is now undergoing a centripetal acceleration toward the center of the circle. The centripetal force is due to friction contact between the tire and the ground.

Step 5. the centripetal force, acting through a lever arm of the mechanical trail, produces a torque that causes the front *wheel to rotate opposite* (counterclockwise) to that of step 2.

Step 6. the front wheel *is turned into alignment* with the frame and rear wheel. The result is a moving bike still leaning toward the right but with the front wheel aligned with the frame.

Step 7. since the front wheel is pointing straight ahead, the bike has lost its centripetal force and now wants to travel in a straight line, tangent to the circular path. This behavior is similar to the rock, whirled in a circle, flying off in a straight line if the string breaks. The tendency to travel in a straight line is inertia.

Step 8. we now have a leaning, moving bicycle with the inertial tendency to 'fly off' in a straight line. However, the bike differs from the rock—the bike is not free to fly through the air. Rather, the wheels are in contact with the ground and friction prevents the bottom of the bike from moving off along the tangent. While the bottom of the bike is 'pinned' to the ground, the bicycle's center of mass is free to move along the tangent. This inertial action of the center of mass results in the bike coming out of the lean and righten to the vertical.

Step 9. as the bike comes out of its lean, it likely continues past the vertical and is now leaning in the opposite (left) direction. We now are back to step 1 with a turn toward the left.

With a rider, steps 1 through 5 make it easier to balance and turn.

Why does a bike need some minimum speed in order to balance?

As long as the bike has sufficient forward speed, the sequence of steps can be repeated many times. Steps 4 and 5 are ineffective with a slow moving bike. Recall that the centripetal force is given by the expression: $F_c = ma_c = m\frac{v^2}{r}$.

At low speeds, there is insufficient centripetal force to straighten out the front wheel and the bike loses its self-stability. *A bike moving too slowly will be very difficult to balance and ride.*

Why do we first countersteer in the opposite direction before beginning a turn?

This phenomenon is identical to the motions required to get the balanced broom to travel in a specific direction. The countersteer puts the bike into a lean. This countersteer is subtle and many riders are not even aware they do it. If you make a determined effort not to countersteer—just try to force a right turn—you will find yourself going left! The forced turn to the right becomes the countersteer for a left hand turn! Our instinctive distaste for falling controls subconscious behavior on the bicycle.

Is it accurate to say a rider turns the bike by leaning her weight rather than turning the handlebar?

A bicycle will travel in the direction the front tire is pointing; if the handlebar is not turned, the bike travels in a straight line. A horizontal turn requires a horizontal force; body weight is a vertical force, it does not point toward the center of turn. It is common, but somewhat erroneous, to say that the body lean turns the bike. Of course, as described in Step 2 above, the body lean does cause the front wheel rotation that now results in a centripetal turning force.

Why does a bike fall over when it encounters street grates or ruts in ice?

Especially hazardous to riders are situations such as street drain grates or ruts in ice. These dangers trap the front wheel and prevent it from turning—the fork is locked! The explanation is simple: if the front wheel cannot turn, the bike cannot balance; therefore, you fall. Trapping of the rear wheel does not cause a loss of balance.

Reference

[1] Kooijman J D G *et al* 2011 A bicycle can be self-stable without gyroscopic or caster effects *Science* **332** 364–342

Appendix A

Bibliography

Knight R P 2008 *Physics for Scientists and Engineers* 2nd edn (San Francisco, CA: Addison Wesley)

Wilson D G 2004 *Bicycling, Science* 3rd edn (Cambridge, MA: MIT Press)

Herlihy D V 2004 *Bicycle* (New Haven, CT: Yale University Press)

Hewitt P G 2009 *Conceptual Physics* 10th edn (San Francisco, CA: Addison-Wesley)

Flannery R P 1989 *Human Physiology* (Philadelphia, PA: Saunders College Publishing)

Adams D R 1996 *Collecting and Restoring Antique Bicycles* (New York: Burgwardt Bicycle Museum)

Velox 1869 *Velocipedes, Bicycles, and Tricycles* (New York: Routledge)

Sharp A 1896 *Bicycles and Tricycles* (Mineola, NY: Dover Publications)

Pratt C E 1879 *The American Bicycler* (Boston, MA: Houghton, Osgood and Company))

Bottomley-Firth J F B 1869 *The Velocipede* (London: Simpkin, Marshall and Company)

Scott R P 1889 *Cycling Art, Energy, Locomotion* (Philadelphia, PA: J B Lippincott Company)

Ward M E *Bicycling for Ladies* (New York: Brentano's)

Oliver S H and Berkebile D H 1974 *Wheels and Wheeling, The Smithsonian Cycle Collection* (Washington DC: Smithsonian Institution Press)

Oliver S H 1953 *Catalog of the Cycle Collection of the Division of Engineering,* United States National Museum, Bulletin 204 (Washington DC: United States Government Printing Office)

Wallis-Tayler A J 1897 *Modern Cycles* (London: Crosby Lockwood and Son)

Garratt H A and Garratt H A 1899 *The Modern Safety Bicycle* (London: Whittaker and Co.)

Hasluck P N 1900 *Cycle Building and Repair* (London: Cassell and Co.)

Berto F 2009 *The Dancing Chain* 3rd edn (San Francisco, CA: Van der Plas Publications)

Bury Viscount and Hiller C L 1887 *Cycling* (London: Longmans)

Roberts D 1991 *Cycling History Myths and Queries The Whole Blooming Truth!* (Hadland)

Ritchie A 1975 *King of the Road - An illustrated History of Cycling* (London: Wildwood House)

doi:10.1088/978-1-6817-4441-4ch13

Appendix B

Common unit conversions

Mass/weight equivalents on the surface of the Earth:

1.0 kg weighs 9.81 newtons(N)

1.0 kg weighs 2.205 lb

1.0 slug weighs 32.2 lb

$g = 9.81\ \mathrm{m/s^2} = 32.2\ \mathrm{ft/s^2}$

Mass:	1.0 slug = 14.6 kg
Force:	1.0 lb = 4.45 N
Distance:	1.0 m = 39.37 inches = 3.28 ft
	1.0 mile = 1.609 km
Speed:	$1.0\ \mathrm{m/s} = 3.60\ \mathrm{km/h} = 2.24\ \mathrm{mi/h}$
	$1.0\ \mathrm{mi/h} = 0.447\ \mathrm{m/s}$
	$88\ \mathrm{ft/s} = 60\ \mathrm{mph}$
Energy:	$1.0\ \mathrm{ft-lb} = 0.3241$ calories
Power:	$1.0\frac{\mathrm{ft-lb}}{\mathrm{s}} = 1.356\frac{\mathrm{N-m}}{\mathrm{s}} = 1.356\frac{\mathrm{joules}}{\mathrm{s}} = 1.356\ \mathrm{watts}$
	$\frac{2000\ \mathrm{Calories}}{\mathrm{day}} = \frac{2\,000\,000\ \mathrm{calories}}{\mathrm{day}} \times \frac{4.184\ \mathrm{joules}}{\mathrm{calorie}} \times \frac{1\ \mathrm{day}}{86{,}400\ \mathrm{s}} = \frac{96.9\ \mathrm{joules}}{\mathrm{s}} = 96.9\ \mathrm{watts}$
Power:	$100\ \mathrm{watts} = \frac{100\ \mathrm{joules}}{\mathrm{s}} \times \frac{1\ \mathrm{calorie}}{4.184\ \mathrm{joules}} \times \frac{3600\ \mathrm{s}}{\mathrm{s}} = 86\,042\frac{\mathrm{calories}}{\mathrm{hour}} = 86\frac{\mathrm{Calories}}{\mathrm{hour}}$
	1.0 horsepower (hp) = 550 ft-lb/s,1 ft-lb/s = 0.00182 hp
	1.0 hp = 746 watts, 1 watt = 0.00134 hp
	1 ft-lb/s = 1.356 watts = 1.165 Calories/hour
	1.0 watt = 0.737 ft-lb/s
	1.0 calorie of heat is equivalent to 4.184 Joules of energy
	1.0 calorie of heat is equivalent to 3.09 ft-lb
	1.0 Calorie = 1.0 kilocalorie = 1000 calories
	one 'serving' of potato chips = 150 Calories = 150 000 calories

doi:10.1088/978-1-6817-4441-4ch14

Appendix C

Trigonometric values

Angle	sin	cos	tan	angle	sin	cos	tan
1	0.0175	0.9998	0.0175	46	0.7193	0.6947	1.0355
2	0.0349	0.9994	0.0349	47	0.7314	0.6820	1.0724
3	0.0523	0.9986	0.0524	48	0.7431	0.6691	1.1106
4	0.0698	0.9976	0.0699	49	0.7547	0.6561	1.1504
5	0.0872	0.9962	0.0875	50	0.7660	0.6428	1.1918
6	0.1045	0.9945	0.1051	51	0.7771	0.6293	1.2349
7	0.1219	0.9925	0.1228	52	0.7880	0.6157	1.2799
8	0.1392	0.9903	0.1405	53	0.7986	0.6018	1.3270
9	0.1564	0.9877	0.1584	54	0.8090	0.5878	1.3764
10	0.1736	0.9848	0.1763	55	0.8192	0.5736	1.4281
11	0.1908	0.9816	0.1944	56	0.8290	0.5592	1.4826
12	0.2079	0.9781	0.2126	57	0.8387	0.5446	1.5399
13	0.2250	0.9744	0.2309	58	0.8480	0.5299	1.6003
14	0.2419	0.9703	0.2493	59	0.8572	0.5150	1.6643
15	0.2588	0.9659	0.2679	60	0.8660	0.5000	1.7321
16	0.2756	0.9613	0.2867	61	0.8746	0.4848	1.8040
17	0.2924	0.9563	0.3057	62	0.8829	0.4695	1.8807
18	0.3090	0.9511	0.3249	63	0.8910	0.4540	1.9626
19	0.3256	0.9455	0.3443	64	0.8988	0.4384	2.0503
20	0.3420	0.9397	0.3640	65	0.9063	0.4226	2.1445
21	0.3584	0.9336	0.3839	66	0.9135	0.4067	2.2460
22	0.3746	0.9272	0.4040	67	0.9205	0.3907	2.3559
23	0.3907	0.9205	0.4245	68	0.9272	0.3746	2.4751
24	0.4067	0.9135	0.4452	69	0.9336	0.3584	2.6051
25	0.4226	0.9063	0.4663	70	0.9397	0.3420	2.7475
26	0.4384	0.8988	0.4877	71	0.9455	0.3256	2.9042
27	0.4540	0.8910	0.5095	72	0.9511	0.3090	3.0777
28	0.4695	0.8829	0.5317	73	0.9563	0.2924	3.2709
29	0.4848	0.8746	0.5543	74	0.9613	0.2756	3.4874
30	0.5000	0.8660	0.5774	75	0.9659	0.2588	3.7321
31	0.5150	0.8572	0.6009	76	0.9703	0.2419	4.0108
32	0.5299	0.8480	0.6249	77	0.9744	0.2250	4.3315
33	0.5446	0.8387	0.6494	78	0.9781	0.2079	4.7046
34	0.5592	0.8290	0.6745	79	0.9816	0.1908	5.1446
35	0.5736	0.8192	0.7002	80	0.9848	0.1736	5.6713
36	0.5878	0.8090	0.7265	81	0.9877	0.1564	6.3138
37	0.6018	0.7986	0.7536	82	0.9903	0.1392	7.1154
38	0.6157	0.7880	0.7813	83	0.9925	0.1219	8.1443
39	0.6293	0.7771	0.8098	84	0.9945	0.1045	9.5144
40	0.6428	0.7660	0.8391	85	0.9962	0.0872	11.4301
41	0.6561	0.7547	0.8693	86	0.9976	0.0698	14.3007
42	0.6691	0.7431	0.9004	87	0.9986	0.0523	19.0811
43	0.6820	0.7314	0.9325	88	0.9994	0.0349	28.6363
44	0.6947	0.7193	0.9657	89	0.9998	0.0175	57.2900
45	0.7071	0.7071	1.0000	90	1.0000	0.0000	undefined

doi:10.1088/978-1-6817-4441-4ch15

www.ingramcontent.com/pod-product-compliance
Lightning Source LLC
LaVergne TN
LVHW080046210926
848481LV00011B/3289
* 9 7 8 1 6 8 1 7 4 4 4 0 7 *